10/7/20

Praise for *Physics and Vertical Causation*

"This short but remarkably rich work by Wolfgang Smith is a *tour de force*, summarizing and synthesizing the seminal texts he has written over the past few decades. Others—Islamic philosophers of Persia in particular—have considered the cosmos, from the invisible to the angelic realms, as an icon and theophany, and have even identified vertical causality with the Divine Will, but these earlier traditional authorities had no knowledge of the sub-corporeal world as a distinct realm—what Dr. Smith calls the 'physical.' In this remarkable synthesis the author finds a cosmic 'home' for the until-now 'homeless' quantum mechanical domain within the world of traditional cosmology, an achievement of the utmost significance not only for physicists seriously interested in the foundations of their science, but also for philosophers and theologians—and, I might add, not only Christian but also Jewish, Muslim, Hindu, Buddhist, as well as of other faiths. May this book receive the global attention it deserves."—SEYYED HOSSEIN NASR, University Professor, George Washington University

"Prof. Smith's metaphysical reading of the cosmic icon by way of its Euclidean construction is on a par with René Guénon's mathematico-metaphysical texts. It even surpasses these in two ways: first, by enabling us to literally *see* what falsifies the Einsteinian conception of 'space-time'; and second, by empowering us to comprehend the cosmological teachings concealed in both the Old and New Testaments through the rediscovery of the cosmic trichotomy."—BRUNO BÉRARD, author of *A Metaphysics of the Christian Mystery*

"As every historian of the field well knows, we owe much of the enterprise of modern science to the studies of metaphysics and natural philosophy pursued by Christian thinkers during the Middle Ages, for it was precisely then that knowledge of the quantitative aspects of nature was not only proven metaphysically possible, technically practicable, and morally desirable, but whole fields of research were launched that resulted three centuries later in the flood of knowledge and mathematical techniques we still teach today in universities. But nowadays we face the opposite question: is mathematical knowledge the only knowledge available to us? Surely not, and no one is better qualified to guide us masterfully through this field than Professor Wolfgang Smith, who has already offered to the public the metaphysical key to understand the subject matter of quantum physics in his now-classic *The Quantum Enigma*. The reader of the present book can expect no less than a study that promises—like its medieval counterparts which opened up the way to present-day mathematical physics—to bear fruit for centuries to come."—RAPHAEL D.M. DE PAOLA, Department of Physics, PUC-Rio

"The importance of this book by physicist and metaphysician Wolfgang Smith lies in the very reasons that will no doubt make it controversial: it challenges certain fundamental features of modern science, namely its epistemological assumptions, its reductive methodologies, and its ideological premises. In so doing, it embraces a broader and more traditional view of science as *Scientia*. This harmonizing, principial knowledge is grounded in an epistemology that respects its ontological roots; in methodologies that go beyond the purely quantitative aspects of mensuration and sensory instrumentality, extending to metaphysical and qualitative intelligibility; and in metaphysical foundations that account for the tripartite structure of reality and explain existence in terms of the principle of vertical causation—which, as Dr. Smith explains, elegantly resolves the 'quantum enigma.' Written lucidly and persuasively, the book summarizes much of his previous writings in the area, and caps a lifetime's work."—M. ALI LAKHANI, Editor, *Sacred Web*

"If we define 'physics' as the science of measurement and distinguish it from the 'philosophy of physics,' we are free to interpret the accumulated data of the former according to our own worldview. New experiments and measurements may offer a bit of guidance regarding which philosophical interpretations of the physical data are more (or less) consistent with the ultimate reality—but for the most part it remains an open question whether we can reach that reality by any one particular method. Professor Smith's latest manuscript, taken together with his earlier work, presents a traditional and ontological interpretation of physics with uncommon quality and competence. What is new about this book is the author's thesis that Einstein's theory of relativity does not conform to such an ontological interpretation, which leads him to conclude that the only tenable position is a geocentric cosmology. Critics who refuse to draw a clear distinction between physics and the philosophy of physics will have nothing of substance to contribute to this discussion."—ALI SEBETCI, Computational Chemical Physicist

PHYSICS AND VERTICAL CAUSATION

The End of Quantum Reality

ALSO BY WOLFGANG SMITH

Cosmos and Transcendence:
Breaking Through the Barrier of Scientistic Belief

Theistic Evolution: *The Teilhardian Heresy*

The Quantum Enigma: *Finding the Hidden Key*

Ancient Wisdom and Modern Misconceptions:
A Critique of Contemporary Scientism

Christian Gnosis:
From St. Paul to Meister Eckhart

Science & Myth: *With a Response
to Stephen Hawking's* The Grand Design

In Quest of Catholicity:
Malachi Martin Responds to Wolfgang Smith
(with Malachi Martin)

Rediscovering the Integral Cosmos:
Physics, Metaphysics, and Vertical Causality
(with Jean Borella)

PHYSICS
and
VERTICAL CAUSATION

The End of Quantum Reality

✦ ✦ ✦

Wolfgang Smith

 Angelico Press

Cover design: Michael Schrauzer

For Thea
Rest in Peace, Beloved

CONTENTS

Preface i

1 The Origin of Quantum Theory 1

2 The Quantum Enigma 7

3 Finding the Hidden Key 13

4 Three Vertical Powers of the Soul 31

Free Will as Vertical Causality [35]—The Verticality of Visual
Perception [37]—The Verticality of Intellect [41]

5 The War on Design 45

The Theoretical Basis of Einsteinian Physics [48]—The
Empirical Argument Contra Einsteinian Physics [62]—
Unmasking "Anthropic Coincidence" [71]

6 The Emergence of the Tripartite Cosmos 77

7 The Primacy of Vertical Causality 93

8 Pondering the Cosmic Icon 101

The Aeviternity of the Spiritual State [106]—The Primacy of
the Intermediary [107]—The Primacy of Time [107]—The
Cyclicity of Time [107]

Postscript 123
Index 127
About the Author 129

PREFACE[1]

THE PRESENT BOOK IS MEANT, IN THE FIRST place, to serve as an introduction to a hitherto unrecognized mode of causation which proves moreover to be ubiquitous: what I refer to, namely, as "vertical causality." The question that immediately presents itself, of course, is how this newly-discovered causality relates to the causality with which physics has been concerned since the days of Sir Isaac Newton, which I shall refer to as "horizontal"; and suffice it to say, by way of a first orientation, that vertical causality does enter into the purview of physics, but in a manner the physicist as such is in principle unable to comprehend. For as we shall come to see, vertical causality—unlike horizontal—is not something quantitative, not something amenable to description in terms of differential equations. At the risk of producing more consternation than enlightenment, one could say that it is a causality that measures but cannot itself be measured. The crucial point is that even though the existence of vertical causation constitutes one of the two keys that render contemporary physics ontologically comprehensible, VC is something by nature invisible to the physicist, and hence proves to be *incurably* philosophical. It pertains moreover to a genre of philosophy which, in the post-Kantian era, has

---

1. Reprinted from "Physics and Vertical Causation" as first published in *Rediscovering the Integral Cosmos* (Brooklyn, NY: Angelico Press, 2018), slightly adapted.

been quite out of fashion: to *metaphysics*. It is thus in a way ironic that this supposedly "outdated" discipline should emerge at the end of the twentieth century as the long sought-after means to understand the latest formulation of physics: that this philosophy should thus live up to its name as constituting indeed a *meta-physics*.

My second objective is to bring into unity the multiple strands pursued in the books I have written over the years, in a way that manifests what may rightfully be termed "the big picture." I take the liberty, moreover, to express myself sometimes in broad sweeps, leaving it to the interested reader to consult this or that earlier work, where a more detailed and documented account of a particular subject is to be found. One has, in the evening of one's life, the luxury to speak freely, and focus on what constitutes the most ultimately profound fact of all. At that point "lesser" facts hardly matter anymore in themselves. What counts in the end is an overview—like the panorama seen from a mountaintop—in which everything finds its rightful place, and "the many" mysteriously unite in that which is incomparably greater than their sum.

✦ ✦ ✦

VERTICAL CAUSALITY MADE ITS APPEARANCE IN THE context of quantum theory, along with the ontological distinction between the physical and the corporeal domains. Not, to be sure, that it was "detected" in the sense of detecting a quantum particle! It is rather implied by virtue of the fact that the resolution of the so-called measurement problem demands as much.

What, then, *is* vertical causality? It needs to be recalled that the causality previously known to physics—which we now qualify as "horizontal"—acts *in time* by way of a temporal process; and as might thus be expected, *vertical* causality is characterized by the fact that it does *not* act in time: one can say that it acts *instantaneously*. How then can one ever "detect" an act of vertical causation: how, in other words, is it possible to conclude that an act of causation was actually *instantaneous* and not just "very fast"? That is where the distinction between ontological domains comes into play: if there exists indeed a *corporeal* domain—the one in which we "live, and move, and have our being"—as distinguished from the *physical* accessed by way of measurement, then it follows that the act of measurement entails a transition from the one to the other: and it is not difficult to see that an *ontological* transition can only be achieved instantaneously.

But whereas vertical causality was discovered in the context of quantum measurement, it proves to be ubiquitous: nothing whatsoever can in fact exist without being "vertically" caused. In particular, it is vertical causality that accounts for the ontological stratification of the cosmos—which the ancients understood so profoundly and present-day civilization fails even to recognize. There is the fact, first of all, that the corporeal world divides into the *mineral, plant, animal* and *anthropic* domains, which prove to be, once again, distinguished *ontologically,* and thus in ways physics as such cannot comprehend—for the very simple reason that, here again, what stands at issue are effects of vertical causality.

To comprehend this hitherto unrecognized mode of causation, we need to understand that the cosmos at large proves to be *ontologically trichotomous*: that even as man himself is made up of *corpus, anima,* and *spiritus,* so is the integral cosmos. Thus, as every major premodern civilization had recognized, there exist two additional ontological strata "above" the corporeal, rendering the cosmos tripartite.[2] There exists moreover a primordial iconic representation of that integral cosmos that proves to be invaluable, consisting quite simply of a circle in which the circumference corresponds to the corporeal world, the center to the spiritual or "celestial" realm, and the interior to the intermediary. What needs above all to be understood—and may indeed be termed the "hidden key"—is that *even as the corporeal domain is subject to the bounds of space and time, the intermediary is subject to time alone while the center is subject to neither of the two bounds.*[3] And so that center—that seeming "point," having neither extension in space nor duration in time, which thus appears to be "the least"— proves to be actually "the greatest of all"[4]: impervious to the constraints of space and the terminations of time, it encompasses in truth every "where" and every "when," and can therefore be identified as the *nunc stans,* the omnipresent "now that stands." Strange as it may seem so long as we picture it as something "far away and high above," that Apex is actually present within every being as its ultimate center. This means that every actual being is endowed with an

---

2. Sanskrit may be the only language with a word for this ontological trichotomy, which is traditionally referred to in India as the *tribhuvāna* or "triple world."

3. I am not aware of any source which enunciates this fundamental principle. As a matter of fact, written sources, both ancient or modern, have precious little to say on the subject of such a "cosmic icon." What we do find are *clues.* A study on this subject would be of great interest.

4. I would note parenthetically that the correspondence here with various allusions in the New Testament—for instance, the parable of the mustard seed—is by no means accidental or adventitious.

PREFACE

ontological "within" centered upon that Apex: it is as if the two centers actually "touch."[5]

To the Thomist let me point out that for every cosmic being, that "meeting point" may be identified with its substantial form. It is needful, therefore, to distinguish between the two centers: the one universal Center (represented by the central point of the cosmic icon), and the other definitive of a cosmic existent. What concerns us is the fact that vertical causality, by virtue of not acting "in time," acts necessarily from a center, and therefore in one of two ways: it may act from the universal Center, in which case that causality coincides with the cosmogenetic Act, or from the center of a particular being, in which case—so far from being cosmogenetic—it is evidently the act of a cosmic agent. And needless to say, there exists a very broad spectrum of such cosmic activity, ranging from the existential act of a pebble to the free and creative acts of man.

✦ ✦ ✦

LET ME, AT THIS JUNCTURE, ASSURE THE READER THAT "the worst" is now over: having plunged ahead into admittedly abstruse and difficult realms—to provide a kind of synoptic overview of the territory we are about to enter—we shall henceforth proceed by clear and simple steps. My aim in this monograph is to provide a readily comprehensible—and exceedingly brief—introduction to the discovery

5. As the reader may note, in the special case of the *anthropos* that "within" coincides with what mystics are wont to call the "heart." It should likewise be noted that the cosmic icon applies not only to the cosmos at large—traditionally termed the *macrocosm*—but likewise to man, the so-called *microcosm*, an analogy which in fact it exemplifies.

and the implications of vertical causality, extending our purview step by step from quantum physics to the cosmos at large: for as we have seen, vertical causality acts in truth from that "*punto dello stelo a cui la prima rota va ditorno*"— from that "*pivot around which the primordial wheel revolves*," to put it in Dante's inimitable words. "*There every where and every when are focused*,"[6] the Poet goes on to say by way of depicting that central and yet ubiquitous *punto dello stelo* wherein the mystery of vertical causation resides.

✦ ✦ ✦

FOLLOWING THE ORIGINAL PUBLICATION OF "Physics and Vertical Causation" in *Rediscovering the Integral Cosmos*,[7] I continued to ponder our iconic circle, what I came to refer to as the "cosmic icon." It occurred to me, moreover, that this figure may well have been viewed in ancient times as the icon *par excellence* of the integral cosmos—a remote conjecture, which however, not long thereafter, gained a measure of credibility from a serendipitous discovery: trying to reconstruct for myself how this "Euclidean" figure might have been viewed within Plato's Academy, I was led to discover what may literally be termed "metaphysical equations": four of them, to be exact, each of which entails a corresponding metaphysical theorem, which in conjunction constitute a metaphysics—a Platonist metaphysics!—of the integral cosmos. I have consequently added these reflections to the present book in the form of chapter 8.

6. *Paradiso* xiii, 10 and xxix, 12.
7. Brooklyn, NY: Angelico Press, 2018.

1

THE ORIGIN
OF QUANTUM THEORY

THE STORY BEGINS IN THE FATEFUL YEAR 1900 when a young physicist named Max Planck decided to investigate the so-called black-body problem. It has always been known that a piece of iron, for example, glows red; yet for some unknown reason it turned out that calculations invariably indicated that it ought to glow blue. Now, to calculate the radiation of a black-body, one needs to relate the kinetic energy E of a vibrating particle to the light-frequency f it emits; and what Planck discovered—serendipitously as it turns out—is that this emission can take place only in "packets" of energy given by hf, where h is a constant subsequently referred to as "the quantum of action" or Planck's constant. Its value has since been established to be 6.626076 times 10 to the power −34 (in standard units); and thus emended, the theory yields the values confirmed by experiment.

As might be expected, this result proved to be utterly incomprehensible to the physics community at the time, and it is safe to say that no one as yet had the ghost of an idea what Planck's discovery presaged. The conviction was rife that physics had attained a state of near-perfection in which only minor problems remained yet to be resolved—which is why the young Planck had in fact been advised by his mentors not to become a physicist, but to pursue instead a career in music! It happens, however, that this state of affairs was about to change.

PHYSICS AND VERTICAL CAUSATION

Since the days of Sir Isaac Newton it had been supposed that matter reduces ultimately to Democritean atoms, and that with the refinement of experimental means these would eventually present themselves as objects to be measured and observed. However, at the very moment when this prospect did materialize, it became apparent that these so-called atoms are not in fact "tiny particles" at all. In place of authentic atoms, what came to light is something that exhibits both particle and wave characteristics, which is to say that it is actually neither a particle nor a wave. Thus, if we do think of it as a particle, we must live with the fact that it can, for example, pass through two slits in a partition at the same time, and can moreover "multilocate" in countless other ways. Nor do we fare any better if we conceive of these entities as "waves," inasmuch as it is now the particle-aspect that does not fit.

One was left thus with something that can no longer be pictured or conceived at all—except possibly in mathematical terms. By the time the "smoke had cleared," physicists were obliged to accept the fact that their near-perfect Newtonian science had, in a sense, vanished into thin air. Of course the theory could still be applied to technology in domains where the deviation from classical behavior is insignificant (which of course covers a very broad range); but that, in any case, is all that remains of the two-century-long Newtonian hegemony.

It was a singularly exciting and challenging time. What was needed was not only a brand new physics that works, but also a new understanding of what physics *is*: that is to say, how it relates to reality. And as we shall come to see, the first of these objectives the physics community was able to achieve rapidly and to perfection, whereas the second they have not been able to attain at all. It thus came about that the most perfect physics the world had ever seen turned out to be "a kind of mystic chant over an unintelligible universe," in Whitehead's telling words.

Let us take a summary look at this new physics. As we have seen, the entities—if indeed that term is still applicable—with which it is concerned exhibit both wave and particle aspects, which implies that in truth they answer to neither designation. It turns out, moreover, that a strict determinism is no longer tenable: somehow the notion of "probability" *must* enter the picture—for it is precisely an "indeterminacy" that allows wave and particle characteristics to co-exist without logical contradiction. The fact is that quantum physics needs both the wave and the particle representation, together with a "principle of indeterminacy" to render them compatible. On top of which it needs one more ingredient: a means, namely, of passing from one to the other. And this is where Planck's constant comes into play: in connecting the *energy* of a vibrating particle to the *frequency* of a wave, it serves as the bridge between the two descriptions.

What was called for, as we have said, was a brand new physics; and amazingly, that transition was accomplished—to perfection!—in a span of twenty-five years. A veritable explosion of genius ensued, such as the world had rarely seen; and in the fateful year 1925—in one giant leap as it were—physics attained what might well be its ultimate form. The discovery was in fact made three times, in terms of three radically dissimilar mathematical structures which later proved to be isomorphic. It was by any count a stellar moment in the history of man's quest to comprehend the universe.

In a way, the greatest genius among the three discoverers was Werner Heisenberg, a 24-year-old physicist and close friend of Niels Bohr, who in July of 1925 sojourned on the desolate isle of Helgoland. Finding himself alone—the sea before him and the sky above—the young physicist was apparently seized by a spirit of "back to the facts." It struck him thus that whereas physicists at large invariably evinced boundless respect for the so-called "hard facts of observation," they seemed rarely to ask themselves what these facts might actually be. They seemed to assume, in particular, that a physical system owns its dynamic attributes—a position or momentum, say—prior to the act of measurement, when in fact this constitutes evidently an unverifiable hypothesis. One may presume that Heisenberg was moved to ask himself whether this may not prove to be indeed the very assumption that renders us incapable of understanding the quantum world! And perhaps, at this juncture, he recalled Lord Kelvin's definition of physics as "the science of measurement"—and realized in a flash that *the mystery of quantum physics resides precisely in the act of measurement itself.*

What we do know is that, abandoning the aforesaid assumption, Heisenberg—in the course of a single day and night—invented a mathematical formalism[1] that enables one to transact the business of physics without assuming that quantum systems own their dynamic attributes. Briefly stated, what a quantum system owns in place of actual dynamic attributes, according to Heisenberg's theory, is an array of probabilities, which could be represented as the elements of an infinite matrix. And unwieldy as the resultant "matrix mechanics" may be, it has now been in direct or indirect use for close to a century and has never yet yielded

---

1. Strictly speaking, he "re-invented" that formalism: what stands at issue is the algebraic theory of matrices, with which the young Heisenberg was not acquainted at the time.

a false result. It would not be unreasonable to suggest that, here at last, physics has attained its ultimate ground in the form of a mathematical science which fully squares with the corresponding facts.

The idea obtrudes that Heisenberg's quantum theory could be something "given" or "discovered" rather than "man-made"; and it appears that Heisenberg himself may have shared that view. As his wife Elisabeth recalls: "With smiling certainty, he once said to me 'I was lucky enough to be allowed once to look over the good Lord's shoulder while he was at work.' That was enough for him, more than enough!"[2]

2. Elisabeth Heisenberg, *Inner Exile: Recollections of a Life with Werner Heisenberg* (Boston: Birkhäuser, 1980), 157.

2

THE QUANTUM ENIGMA

Y ET IT APPEARS THAT THE CONSUMMATE
perfection of quantum theory comes at a price: for
it happens that no one seems to have so much as the
slightest notion what in plain fact it means—whether, for
example, there actually exists a "quantum world" or not.
Now it seems that physicists at large have not been unduly
disturbed by this state of affairs. Most seem content to vacil-
late between the pre-quantum outlook on the one hand and
some suitably reified picture of the quantum realm. They
appear for the most part not even to realize that something
is seriously amiss, and that in fact they are pendulating
between two contradictory worldviews. In the upper eche-
lons of the physics community, on the other hand, efforts
began almost immediately to render the marvelous new
physics intelligible as well. For that subclass of physicists
who are more than technicians, predicting the outcome of
experiments was not enough; they wanted also to understand what physics entails regarding the actual composition
of the world, and found it intolerable that the new science
seemed not to fit the prevailing worldview. A debate concerning these deeper issues began almost immediately at the
Solvay Conference of 1927 in the form of the famous Bohr-
Einstein exchange, and for the past ninety years, to be precise, physicists with a philosophic bent have proposed and
counter-proposed the most extraordinary notions in an

effort to resolve the persisting riddle, *with no resolution yet in sight*. It appears that Richard Feynman may well have hit the nail on the head with his apodictic declaration "*No one understands quantum theory.*" Many do, of course, understand the theory on a technical plane: the "*no one*" applies—not evidently when one compares mathematical solutions with corresponding measurements—but the moment one asks how a red apple in our hand relates to protons and electrons.

As concerns the manifold proposals put forth by physicists in the interminable quest to resolve the "quantum reality" conundrum, it needs first of all to be noted that, in the final count, each without exception falls short of the mark. On the whole these proposals strike the "unprogrammed" observer as ranging—let me speak plainly—from the bizarre to the outright ridiculous, and none more so, I would add, than the so-called "many-worlds" approach which seeks to rescue determinism by stipulating that every possible outcome of every measurement is in fact realized, *howbeit in a different universe!* So too mention might be made of the so-called "quantum logic" approach, based upon the remarkable premise that ordinary logic ceases to apply in the quantum realm. The very absurdity of such proposals—combined with the fact that they originated in universities and institutes for advanced study—serves to underscore the difficulty and indeed profundity of the quantum enigma.

I shall argue that it is the so-called Copenhagen interpretation, originally conceived by Niels Bohr, that *beyond all doubt* holds precedence over all competing views regarding the nature of quantum reality by virtue of its pivotal tenet, which affirms that *a quantum system does not own its dynamic attributes* (such as position or momentum). As previously noted, this was in essence the "back to the facts" recognition which inspired Heisenberg, on that fateful day in 1925, to achieve his monumental breakthrough. Meanwhile that stipulation was confirmed, in 1932, by a Hungarian mathematician named John von Neumann, in a startling revelation which appeared to settle the issue beyond all doubt. What von Neumann accomplished breaks into two parts: first, he axiomatized the principles of quantum physics, thereby putting the discipline upon a rigorous mathematical foundation, following which he showed that the Copenhagenist claim of no "pre-measured" dynamic attributes can now be established as a mathematical theorem. Thus, if we define an *ordinary* object as one that owns its dynamic attributes, the theorem states quite simply that *there are no* ordinary objects in the quantum realm.

But whereas it appeared at this point that the matter had now been settled once and for all, it turns out that additional feats of genius were in the offing, which would, once again, upset the *status quo*. For it turns out that von Neumann had neglected to spell out a certain condition that had hitherto been assumed as a matter of course, but which proves ultimately to be not only unwarranted but indeed untenable. Which brings us to John Stewart Bell, the physicist who, in 1964, in the course of an exacting study of von Neumann's proof, identified that mysterious condition. What von Neumann had tacitly assumed—and what later came to be known as the condition of *locality*—is that quantum particles can interact only via forces which propagate no faster than the speed of light. It thus turns out that what von Neumann had actually established as a theorem of quantum mechanics is that *local* objects do not own their dynamic attributes: which is to say that *ordinary objects must be nonlocal.*

PHYSICS AND VERTICAL CAUSATION

✦ ✦ ✦

FROM THE OUTSET THE EPICENTER OF THE QUANTUM REALITY debate was defined by the Bohr-Einstein exchange. Quantum theory had scarcely seen the light of day when the renowned Albert Einstein stepped forth to challenge Niels Bohr in a bid to overthrow the new physics, or at the very least, reduce it to a mere approximation "beneath" which a basically Einsteinian physics reigns supreme. What troubled Einstein the most in regard to quantum theory, it seems, is the loss of a Newtonian or "absolute" determinism, that is to say, its replacement by an incurably *probabilistic* physics; as he famously put it: "*God does not play dice.*" Moreover, inasmuch as an absolute determinism demands "ordinary" objects, Einstein was adamantly opposed to the central tenet of the Copenhagenist position. He consequently proposed various arguments and a plethora of ingenious experiments designed to reinstate ordinary objects in a bid to disprove the quantum-mechanical indeterminism. The irony is that one of these—the so-called Einstein-Podolski-Rosen or EPR experiment—has led eventually to a result henceforth known as Bell's theorem, which constitutes arguably an irrefutable *disproof* of Einstein's contention.

To indicate briefly what stands at issue, let us consider a somewhat simplified version of the relevant EPR experiment. Consider two photons in a so-called "state of parallel polarization," emanating in opposite directions from a source O to points A and B, which can be as widely separated as we wish. Whereas neither particle possesses a definite polarization initially, this entails that once a polarization has been established by an act of measurement at A (let us say), the same polarization will be found in the photon at B: it is as if the measurement at A affects the state of the photon at B *instantaneously.* Thus, if one can still speak of "interaction" at all, what confronts us here is indeed a *nonlocal* interaction: the kind von Neumann had inadvertently left out of account. And so his famous theorem, as we have noted before, does not in fact rule out *ordinary* objects *per se*—as everyone had initially believed—but only *local* ordinary objects: the kind that interact only via forces carried by fields.

This brings us finally to what has come to be known as Bell's theorem, an epochal result the Irish physicist discovered in the wake of his inquiry into von Neumann's proof. By means of an argument inspired by the EPR setup, and which, though ingenious in the extreme, turns out to be relatively uncomplicated, Bell proved—to everyone's utter amazement!—that actually *there are no local objects*: in a word: *reality is nonlocal.* Now the implications of this discovery prove to be unimaginably far-reaching, and for the most part remain yet to be explored. Meanwhile, one would be hard pressed not to concur with Berkeley physicist Henry Stapp when he declares Bell's theorem to be "the most profound discovery of science."[1]

The question now becomes what Bell's theorem has to say regarding Einstein's view of quantum theory. It is true that *ordinary* objects have not in fact been ruled out, as von Neumann's theorem had initially seemed to imply. But meanwhile something no less unacceptable to Einsteinians has come to light: the fact, namely, that *physical objects*— including the *ordinary* kind, if such do in fact exist—*must be nonlocal*: have the capacity, that is, to communicate with

1. "Bell's Theorem and World Process," *Il Nuovo Cimento*, 40b (1977), 271. For a readable explanation of Bell's proof we refer to Nick Herbert, *Quantum Reality* (New York: Doubleday, 1985), 211–31.

other such objects *instantaneously*, a possibility which the very axioms of relativistic physics stringently exclude. The irony is that it was the EPR set-up—which Einstein himself had conceived in the expectation that it would disprove the claims of quantum theory—that had enabled Bell to prove his "anti-Einsteinian" theorem.

✦ ✦ ✦

GETTING BACK TO THE INTERPRETATION OF QUANTUM theory: the superiority of the Copenhagen approach, I would argue, derives from its care not to overstep the bounds of what we actually know: only in the wake of such an over-reach, I surmise, does one experience a need for something as weird as a "multiverse" or a so-called "quantum logic." As Niels Bohr reminds us, even to speak of a "quantum world" is to overstep what we actually know: a "quantum description" is all we can legitimately claim. And that description is moreover geared to the business of physics: beyond this, its intended and rightful application, no one indeed "understands quantum theory"! It was the philosopher Alfred North Whitehead—and not Niels Bohr—who lamented the fact that physics has thus been reduced to "a mystic chant over an unintelligible universe." How quantum physics could be more than a "mystic chant"—that problem remained yet to be resolved.

3

FINDING
THE HIDDEN KEY

NOT LONG AFTER I HAD BEGUN TO PURUSE the quantum reality literature, I was struck by the fact that everyone seemed implicitly to presuppose a major philosophic postulate, which at the very least could be characterized as "dubious." Whereas, as a rule, assumptions of even the most seemingly innocuous kind were sought out meticulously and subjected to exacting scrutiny by one or another of the quantum-reality theorists—even the hitherto sacrosanct principles of logic!—I was amazed to find that the Cartesian premises, which entered the scientific mainstream by way of Newton's *Principia*, had apparently remained undetected, and in any case unchallenged by the investigators. What stands at issue in this philosophic Ansatz is a splitting of the real into two mutually exclusive compartments: an external world comprised of so-called *res extensae* or "extended entities," and an internal and subjective domain consisting of *res cogitantes* or "thinking entities." And even though this Cartesian "bifurcation" has been often enough called into question by philosophers of rank, and Alfred North Whitehead in particular has chided the scientific community repeatedly for its adhesion to what he termed "the fallacy of misplaced concreteness," it appears these strictures have invariably fallen upon deaf ears. Whitehead himself, moreover, offers at least one explanation as to why the Cartesian philosophy has thus become *de facto* sacrosanct:

> In the first place, we must note its astounding efficiency as a system of concepts for the organization of scientific research. . . . Every university in the world organizes itself in accordance with it. No alternative system of organizing the pursuit of scientific truth has been suggested. It is not only reigning, but it is without a rival. And yet—it is quite unbelievable.[1]

One might add that these words were written prior to 1925, the year when everything pertaining to physics changed—or should by right have changed. When it comes to "the system without rival," however, what came to pass was a classic case of "pouring new wine into old bottles": Cartesian bottles, to be exact.

The main exception among the reigning authorities, it seems, was Werner Heisenberg, who did at times voice doubts concerning the Cartesian postulates, and lamented that "today in the physics of elementary particles, good physics is unconsciously being spoiled by bad philosophy."[2] Now, it seemed to me that what was actually being "spoiled" by that "bad philosophy" was not in fact the "good physics" itself, but the *ontological interpretation* of that "good physics," which is something else entirely. And so I arrived at the surmise that it must be, at bottom, the Cartesian partition of the real into *res extensae* and *res cogitantes* that accounts for the fact that, to this day, "*no one understands quantum theory.*"

The first thing that needed to be done to break the Cartesian stranglehold was to see how the postulate of bifurcation stands on philosophic ground: whether, in other words, it is

---

1. *Science and the Modern World* (New York: Macmillan, 1925), 54.
2. *Encounters with Einstein* (Princeton University Press, 1983), 81.

well founded. What struck me was the fact that, in essence, Descartes had simply reinstated the Democritean atomism—the notion that *"by convention there exist color, the sweet, and the bitter, but in reality only atoms and the void"*[3] as the celebrated fragment has it—a doctrine which the major schools of Greek philosophy came in time to view as heterodox. The key issue, it seemed to me, is to grasp how sense perception—and *visual* perception above all—can transcend the subjective realm of *res cogitantes* so as to perceive, not a mere image, but indeed the external object itself: the world of *res extensae*. And so, in the first chapter of *The Quantum Enigma*, I dealt with this question as best I could—unaware of the fact that, not long before, a cognitive psychologist by the name of James Gibson had in effect resolved this issue on rigorous scientific ground. To be precise: having discovered, by means of key experiments, that the prevailing "visual image" theory of perception proves to be untenable, and having subsequently spent three decades in quest of a new paradigm, Gibson arrived at what he terms "the ecological theory of visual perception," which—incredible as it may seem—*falsifies* the Cartesian premise *on scientific grounds*, and re-establishes what amounts to the pre-Cartesian realism.[4] We will recur to this crucial issue in the next chapter.

Given then that we do perceive the external world—that the grass is actually green and the red apple in my hand is not after all a *res cogitans*—given this fundamental premise, I posed the question whether it is possible to interpret phys-

3. Hermann Diels, *Fragmente der Vorsokratiker* (Dublin: Wiedemann, 1969), vol. II, 168.

4. James Gibson, *The Ecological Approach to Visual Perception* (Hillsdale, NJ: Erlbaum Publications, 1986). I have discussed Gibson's theory at length in *Science and Myth* (Tacoma, WA: Angelico Press/Sophia Perennis, 2012), ch. 4.

ics *per se*, based on its inherent *modus operandi*, in non-bifurcationist terms. And let it be noted at once: if such be indeed the case, it follows that *the non-bifurcationist interpretation of physics cannot be rejected on scientific grounds.* Or to put it another way: if indeed it is possible to transact the business of physics on a non-bifurcationist basis, then— contrary to the prevailing belief—Cartesian bifurcation is bereft of scientific support. *The contemporary Weltanschauung—which implicitly assumes bifurcation to be a scientific fact—has then been disproved.*

Such was the plan; and its execution proved to be surprisingly uncomplicated. The first and crucial step was to distinguish *ontologically* between *corporeal* and *physical* objects, based upon the fact that these answer to fundamentally different ways of knowing: to direct perception on the one hand, and to the *modus operandi* of physics on the other. *Corporeal* objects, then, are the kind we perceive, whereas things *physical* are the kind we know by empirical means, i.e., by way of *mensuration.* And these categories define two distinct ontological planes: the *corporeal*, which is accessed by way of sense perception, and the *physical*, which has come into view through the discoveries of physics.[5] I would add that I regard the ontological distinction between these two planes to be crucial for the resolution of the quantum enigma: that in fact a philosophic comprehension of the quantum realm hinges precisely upon that recognition.

---

5. There is reason to believe that, to some extent at least, the physical universe is actually "constructed" by the intervention of the physicist, which is the reason John Wheeler refers to it as "the participatory universe," and why Heisenberg states that physics deals, not with Nature as such, but with "our relations to Nature." I have dealt with this issue at length in "Eddington and the Primacy of the Corporeal." See *Ancient Wisdom and Modern Misconceptions* (Kettering, OH: Angelico Press/ Sophia Perennis, 2015), ch. 2.

But let us continue: inasmuch as we find ourselves thus confronted by two distinct ontological planes, the economy of physics demands the existence of a bridge which enables us to pass from one to the other, in the absence of which there evidently could be no physical science at all. Now it turns out that there exists one and only one such bridge, which unbeknownst to the scientific community has been in constant use since the days of Newton: what then is that "invisible" connecting link? It is defined by the function S which to every corporeal object X assigns the corresponding *physical* object SX, which is simply *X as conceived by the physicist*. But let us not fail to note that X and SX prove to be as different as night and day: the corporeal object X has color, for example, and owns a host of other *qualitative* attributes (failing which it could not be perceived), not one of which pertains to SX by virtue of the fact that the latter is described exclusively in mathematical terms. It is moreover to be noted that the "bridge" is crossed from X to SX by the theoretician, and again from SX to X by the experimentalist in the act of measurement. *The economy of physics hinges thus upon two "crossings" of that bridge*: a theoretical transition from X to SX, complemented by an empirical return from SX to X.

These, then, are the basic conceptions that enable a non-bifurcationist interpretation of physics; and let me emphasize that what is thus specified is not a particular *kind* of physics, but physics *per se* as defined by its founding *modus operandi*. That interpretation is consequently immune from contradiction on scientific grounds: it simply brings to light what exactly physics *is*, and what *kind* of knowledge it supplies. The primary reference of physics, it thus turns out, is not in fact to the corporeal world, but to the physical domain, which proves moreover to be actualized by its own *modus operandi.* The reason why physics does nonetheless have something to say regarding the *corporeal* realm resides in the fact that the latter derives its *quantitative* content from the *physical.* One may, of course, go on to ask all manner of questions concerning the nature and origin of that "subcorporeal" domain, its relation to other ontological strata, and so forth: the point, however, is that these are *philosophic* questions which physics as such is neither able nor in any way obliged to answer.

✦ ✦ ✦

WE NEED NOW TO ASK WHAT IT IS THAT DISTINGUISHES the corporeal from the physical; and the first thing to note is that this, too, is a question physical science as such cannot pose, let alone answer. *Physics has eyes only for the physical,* period; and one might add that inestimable harm to civilization at large has resulted from the fact that this inherently obvious fact has as a rule been negated by the presiding elite. The almost universal tendency on the part of physicists to conflate the physical and the corporeal domains turns out thus to constitute a category error resulting from a failure to comprehend the *modus operandi* of physics in ontological terms.

What in truth distinguishes corporeal entities from the physical is the fact that they exist. This may of course come as a shock to the public at large inasmuch as the prevailing worldview affirms in effect the very opposite. In the name of science an uncanny deception has been imposed upon humanity, which—quite literally—*stands the world on its head.* Yet the fact remains that *the physical,* properly so called—so far from coinciding with the corporeal—constitutes in truth a *sub-existential* domain. And this should in fact come as no surprise if only one recalls that Heisenberg himself has situated the so-called elementary particles ontologically "just in the middle between possibility and reality," and has pointed out that as such they are in fact "reminiscent of Aristotelian *potentiae*."[6] We need thus to ask the crucial question: what is it, then, that *actualizes* these *potentiae*? And as is so often the case when, at long last, the right question is posed, the answer stares us in the face! *Quantum particles are "actualized" precisely in the act of measurement*, and thus *on the corporeal plane*: in the state of a *corporeal* instrument, to be exact.

The key to the quantum enigma is thus to be found in Lord Kelvin's conception of physics as "*the science of measurement*," which is to say that the ultimate object of physics is neither the physical as such, much less the corporeal, but the transition, precisely, from the former to the latter in the act of measurement. A so-called quantum particle, thus, is not *actually* a particle—does not, properly speaking, *exist!*—until it interacts with a corporeal instrument of measurement or detection: "beneath" the corporeal plane all is potency, what Heisenberg refers to as "Aristotelian *potentiae*." In the final count, one is forced to admit that the *physical* universe, properly so called, constitutes, properly speaking, a *sub-existential* domain, which in an unimaginably subtle yet absolutely precise sense *underlies* the corporeal world, and in fact determines its *quantitative* attributes.

✦ ✦ ✦

WE SHOULD REMIND OURSELVES, AT THIS JUNCTURE, THAT the notion of an ontological realm or stratum "*beneath*" the

6. *Physics and Philosophy* (New York: Harper & Row, 1962), 41.

corporeal proves to be integral to our metaphysical heritage. It springs in fact from the seminal recognition that corporeal being entails, not one, but *two* fundamental principles: something called *hyle* or *materia*, plus *morphe* or *form*, to put it in Aristotelian terms. Everything in creation hinges upon these two complementary principles: the *paternal*, exemplified by *form*, and the *maternal* corresponding to *materia* (it seems our very language testifies to this fact). The doctrine known as *hylomorphism* proves thus to be—not the mere invention of Aristotle—but the expression of a universal truth, which in one form or another constitutes in fact the *sine qua non* of every sound ontology.

It will be expedient at this point to observe that this foundational duality of *form* and *matter* has been conceived *iconically*—from time immemorial—as a *vertical* distinction, which is to say that one conceives of *morphe* "pictorially" as situated "above" *materia*, a step which defines a "vertical" axis, a cosmic "up" and "down," a "high" and "low." It is hard to describe or explain what actually stands at issue here because the idea is incurably metaphysical, and yet so very "cosmic," so very "pictorial" one could even say. Yet the fact is that the authentic concept of "verticality" is both factual and normative, a universal compass one might say, accessible to mankind. I therefore ask the reader's indulgence: bear with me when I speak of "verticality"—of "high" and "low"—in so many contexts, while insisting that these are not merely "subjective" conceptions. More than that is in play: what stands at issue, in the final count, *is nothing less than the metaphysical foundation of the world*, which resides in the *morphe/hyle* dichotomy.[7]

---

7. I cannot refrain from pointing out that it is precisely the loss of this metaphysical recognition that has led to the well-nigh universal relativism

In terms of this metaphysical and archetypal symbolism we can "picture" the integral cosmos as a hierarchy of "horizontal" planes;[8] which brings us to the question: where then—on what "level"—are we to situate this, our *corporeal* world? And contrary to contemporary expectation, metaphysical tradition answers with one voice: *on the very lowest tier!* According to the simplest and if you will primary representation, we arrive thus at three tiers—corresponding, as we have noted before, to the *corpus-anima-spiritus* ternary. The corporeal plane of the integral cosmos corresponds thus to the "corporeal" component of man. This is where "existence," properly so called, comes to an end—or begins, if you look at it the other way. *Below* the corporeal world there is "pure *materia*," so-called *materia prima*, conceived as a "pure receptivity" which as such does not actually exist.

Now, all this having been said, we can return to the quantum enigma and state, quite simply, that the physical universe is situated *below* the corporeal, or more precisely: "between" the corporeal and *materia prima*. One might add that this prime or first "*materia*" corresponds to the so-called "waters" over which "*the spirit of God*" was said to "move"—or even more suggestively, perhaps, to the *Chaos* which, according to Hesiod, "*was in the beginning.*" These in any case are the metaphysical categories, come down to us from remote antiquity, in terms of which one can "situate" the newly-discovered quantum world. To do so in terms of

in regard to values and norms of every "non-quantitative" kind, be it in the sphere of art, morality, or, of course, religion. The idea that "values" are ultimately founded upon *truth* is perhaps the contra-modern recognition most desperately needed in our day.

8. Or as concentric rings, in accordance with what we have termed, in the preface, the "cosmic icon."

concepts less ancient and less venerable would prove, I believe, to be misconceived.

I would add—for those who have "eyes to see"—that these
ancient and perhaps pre-historic cosmological conceptions
carry not only an archetypal, but a *mystical* sense impenetra-
ble to the modern mind. Yet even so we can draw inspiration
from that perennial source, and *need in fact to do so* if ever
we are to "understand quantum theory." It might be worth
pointing out that Hesiod's allusion to a sub-existential
Chaos—which can after all be conceived as a kind of co-
presence of contradictory elements—may not be irrelevant
to the metaphysics of quantum theory, which does after all
entail a "co-presence" of mutually exclusive states[9] by virtue
of the superposition principle. Could it be, then, that the
quantum realm rests upon or embodies that Hesiodian
Chaos? What in any case has been established beyond doubt
through the resolution of the quantum enigma is that the
quantum world—or more precisely, what I term the *physical
universe*—proves to be a sub-existential domain, situated
ontologically between *prima materia* and the corporeal
plane.

✦ ✦ ✦

HAVING THUS "SITUATED" THE PHYSICAL OR QUANTUM
world, it is crucial to note that in this subcorporeal realm the
concept of "substance" ceases to apply: as Arthur Eddington
was quick to recognize: "*the concept of substance*" has indeed
"*disappeared from fundamental physics.*"[10] The problem,

---

9. As distinguished from an actual coexistence, which would be antin-
omous.

10. *The Philosophy of Physical Science* (Cambridge University Press,
1949), 110.

FINDING THE HIDDEN KEY

however, is that so far hardly anyone, even within the academic ranks, appears to have grasped the point: to this day the tendency to reify the quantum world *at the expense of the corporeal*—as a mark of up-to-date enlightenment, no less!—is rife across the land. I recall with consternation, but little surprise, the words of a top Vatican spokesman, who announced that the concept of "transubstantiation" is no longer tenable "because physics has proved that there is no such thing as *substance*." It is time, I say, to put an end to this catastrophic confusion, imposed upon present-day humanity by the reigning *periti*: high time, indeed, to recover our collective sanity!

To this end let it be recalled, once again, that this intellectual—and spiritual—bedlam is based squarely upon Cartesian bifurcation, which to this day serves as the bedrock of our supposedly "scientific" Weltanschauung. It is astounding to observe what a veritable stranglehold this unlikely and indeed untenable premise exerts to this day upon our post-Newtonian civilization: how the most basic tenets our science is said to have "proved" rest squarely upon that illusory foundation.

What is it then, of which Cartesian bifurcation has in effect deprived the external world, that mankind so desperately needs? What could it be, in the absence of which we lose our bearing and become progressively dehumanized? The answer to this vital question is that the precious ingredient is given to us in the apprehension of what we have termed *verticality*: in the fact, namely, that the cosmos presentifies not only entities, but *values*, that it speaks to us not only of "things," but of *beauty* and *goodness*—and ultimately, as Plato informs us, of the Beautiful and the Good itself. We need to remind ourselves thus of the categorical distinction between *qualities* and *quantities*, which proves to be immeasurably profound: for whereas *quantities* derive in truth "from below"—in keeping with the Scholastic dictum *"numerus stat ex parte materiae"*—it can in truth be said that *qualities* stem "from above," that in fact *they transmit the light of supernal essences into this nether world.*

It thus becomes evident—on the basis of what might truly be termed *the perennial wisdom of mankind*—that in banishing the *qualitative* content of the corporeal world by relegating qualities *per se* to a subjective realm of so-called *res cogitantes,* Descartes has in effect cast out the very *essence* of the world in which we live, and move, and have our being. The crucial and almost universally undiscerned fact is that the Cartesian reduction of the corporeal world to "matter"—the denial, thus, of its "formal" component, its inherent *morphe*—has seemingly emptied the world of everything that answers to the higher cravings of the human heart. And of all that has thus been forfeited, *the loss of the sacred* is beyond doubt the most tragic of all: for that proves to be the privation we cannot ultimately survive.

✦ ✦ ✦

MEANWHILE—IN THE WAKE OF THE CARTESIAN DISCLO-sure—it seemed to the dazzled progeny of the Enlightenment that every new triumph of Newtonian physics constituted yet another proof of that *non plus ultra* Weltanschauung. And so the victorious march continued right up to the twentieth century, at which point something utterly unexpected occurred: at the precise moment when that all-conquering physics finally attained to its own ground by ridding itself of its last non-quantitative vestiges, it came to pass that the very idea of "substance"—and thus of *being*—had to be jettisoned: in the so-called "quantum world" there

FINDING THE HIDDEN KEY

is no substance, no *being* whatsoever! And from that point onwards, as we have seen, physicists have been confronted by the daunting task of constructing a universe out of "things" that do not actually exist: no wonder that labor continues up to the present day!

Given that Descartes propounded his fateful postulate to prepare the way for a mathematical physics of unlimited scope, it is ironic that the very science to which that Ansatz gave rise has led in the end to a veritable *reductio ad absurdum* which in effect disqualifies the Cartesian worldview itself. Thus it came to pass in 1897—just when the quest for *res extensae* in the form of Democritean atoms seemed finally to have attained its end in J. J. Thomson's discovery of the electron—that at this very moment the long-sought quarry mysteriously eluded our grasp. And by 1925, following one of the most intense periods of intellectual endeavor in human history, the verdict was in: *there is actually no such thing as a "fundamental particle,"* no such thing as a *res extensa* at all! One might add that the non-existence of *res extensae* could indeed have been foreseen by anyone conversant with the traditions of metaphysics; what proves to have been humanly unforeseeable, on the other hand, is the discovery of a subcorporeal domain, made up of entities "midway between being and nonbeing," from whence the *quantitative* attributes of the corporeal universe are derived, along with a mathematical physics of unimaginable accuracy descriptive of this realm, which gives rise moreover to technological wonders that not only "*could deceive even the elect,*"[11] but arguably have done so.

11. Matt. 24:24.

PHYSICS AND VERTICAL CAUSATION

✦ ✦ ✦

GETTING BACK TO THE QUANTUM ENIGMA: IT NEEDS NOW to be pointed out that what I have termed "the rediscovery of the corporeal world" proves to be but the first step in the resolution of the quantum quandary. What is likewise required, it turns out—in addition to a supra-physical world—is a supra-physical mode of causation. Thus, in addition to the modes of causality with which physical science is conversant—which operate by way of a temporal process that may be deterministic, random or stochastic, and which we shall henceforth refer to as "*horizontal*"— there exists also what I term a *vertical* mode of causality, which does *not* operate in time, but acts *instantaneously* and therefore, as it were, "above time." It turns out, moreover, that this hitherto unrecognized causality plays a crucial role in the economy of physics, and thus constitutes yet another reason why "nobody understands quantum theory."

As the ontological distinction between the *physical* and the *corporeal* might lead one to surmise, it is precisely in the act of measurement that *vertical* causation comes perforce into play. For it stands to reason that *horizontal causation cannot act from one ontological plane to another*—given that such an effect must be *instantaneous*. Inasmuch, therefore, as the act of measurement does entail an interaction between the *physical* system and a *corporeal* instrument, it cannot be attributed to horizontal causation: the matter turns out to be as simple as that! The fact is that, at the instant of measurement, the evolution of the physical system, as described say by the Schrödinger wave equation, is interrupted—the Schrödinger equation is "re-initialized," as physicists say—an event for which there is no *physical* explanation. In fact, there *cannot* be: what confronts us here proves incontrovertibly to constitute an effect of *vertical* causation, an act which affects both the measuring instrument and the physical system *instantaneously*. It should be noted, moreover, that this causation "emanates," not from the physical, but perforce from the corporeal side, inasmuch as it interrupts—or "overrides"—the "Schrödinger causality" indigenous to the physical plane.

I would point out, moreover, that this settles the long-debated question whether the postulate of a perfect or so-called "Laplacian" determinism has or has not been disqualified by quantum theory: it can in fact be seen that it has, though not on purely mathematical grounds. For insofar as the act of measurement entails an ontological transition from the physical to the corporeal domain, it cannot be fully described in mathematical terms: no differential equation, evidently, can determine the outcome of an *instantaneous* act! And let us note that this interdict refers in particular to the differential equations of the de Broglie-Bohm theory, which despite their marvelous elegance and rigor of mathematical derivation still do not apply to an act that does not transpire in time. The ineluctable verdict, thus, is that the notion of a Laplacian or "mathematical" determinism has now been rigorously disqualified, *and no "hidden variables" theory can alter this fact.*

Getting back to vertical causation: one recognition, as is so often the case, leads to another. Even as the rediscovery of the corporeal domain has led to the identification of vertical causality in the act of measurement, it now becomes apparent that other acts of VC are occurring ubiquitously, beginning with the fact that every corporeal object X acts "vertically" upon the corresponding physical object SX—which is evidently the reason why cricket balls don't multilocate and cats cannot be both dead and alive.[12] The point is that the physical object SX is invariably constrained by its corporeal counterpart to exclude superpositions incompatible with the corporeal nature of X. And thus another long-standing conundrum of quantum theory has been resolved.

The most profoundly significant fact of all, however, is that *the effect of vertical causation emanating from a corporeal object X is by no means limited to the immediate vicinity of X, but can in principle encompass all of space!* And herein indeed resides the wonder of Bell's theorem,[13] which has caused such consternation among physicists. Let me recall the salient fact as illustrated by the example of the twin photons in a state of "parallel polarization," traveling respectively to points A and B, perhaps millions of miles apart. The fact is that a measurement which determines the polarization of the photon at A *instantly* determines the polarization of its twin at B as well. Now, what shocks physicists in this scenario is the fact that *horizontal* causality is categorically incapable of accounting for this effect—which however signifies simply that the prodigy constitutes perforce an effect of *vertical* causation. Bell's theorem affirming the non-existence of "local" objects thus entails not only the "ubiquity" of vertical causation, but its well-nigh miraculous efficacy as well: for not only does VC act instantaneously, but *its effect is not diminished by spatial separation.*

From a metaphysical point of view, that well-nigh miraculous efficacy is implied by the fact that vertical causality operates *instantaneously*; for as we have noted at the outset, this situates the origin of that causality at the highest level of

---

12. I am referring to the so-called "Schrödinger paradox." See *Ancient Wisdom and Modern Misconceptions*, ch. 1.

13. See chapter 2, 11–12.

the cosmic hierarchy: at the Apex of the integral cosmos, one can say, which transcends not only the bound of time, but consequently the bound of space as well. Vertical causality proves thus to be not only "higher" but incomparably more powerful than the modes of causality hitherto known to physics, which actually stem from the opposite direction: that is, from the subcorporeal domain.[14]

14. It must not be forgotten, however, that even though horizontal causality derives from a subcorporeal plane, it is yet "activated" by vertical causality, as is clearly evident in the instance of quantum measurement.

4

THREE VERTICAL
POWERS OF THE SOUL

The soul is partly in eternity, and partly in time.
PLATO

THIS BRINGS TO A CLOSE OUR CYCLE OF
reflections centered upon quantum physics. As we
have come to see, the resolution of the quantum
enigma hinges upon two primary distinctions: the *ontological* discernment between the *corporeal* and the *physical*
domains, and the *etiological* between *horizontal* and *vertical*
causation. But whereas the notion of corporeal being constitutes a basic conception of traditional philosophy, it
appears that the existence of vertical causality has remained
essentially unrecognized and unexplored.

It is time, I believe, to remedy this neglect: the present crisis demands no less. Having been subjected to what might
be termed "the tyranny of horizontal causation" since the
onset of the Enlightenment, it is high time to realize that all
the causalities known to physics prove in fact to be *secondary*, with a sphere of operation restricted to the lower
extremities of the integral cosmos, whereas the primary
mode of causation—whose sphere of action extends
through the length and breadth of the cosmos, from its
Apex down to a particle of dust—proves to be inexorably
vertical. The moment has arrived, it seems to me, for this

decisive fact to be brought into the light of day: the defeat of scientistic materialism—the restoration of our collective sanity!—appears to demand nothing less.

Given that, among natural functions, human intelligence may be the most closely connected to "vertical" causality, let us now shift our attention, from fundamental physics to the opposite end of the *scala naturae*, to investigate the active and cognitive faculties of mankind, in the expectation that these will prove in fact to be *vertical*, that is to say, supratemporal in their mode of action. It turns out that not only God, but man too has a certain "access" to the *nunc stans*: the elusive "now that stands," which, as we have noted, constitutes the central Point and Apex of the integral cosmos.

✦ ✦ ✦

INASMUCH AS WE SHALL BE DEALING WITH "POWERS OF the soul," it is needful, first of all, to consider—following centuries of oblivion—what exactly is meant by a "soul." Recalling the Aristotelian distinction between *morphe* and *hyle* ("form" and "matter"), let us note that a corporeal entity is what it is by virtue of its "formal" component, known as a "substantial form." Thus, in the absence of a substantial form, that putative entity *has no* "what," which of course entails that like the so-called fundamental particles of quantum physics, it does not actually exist as an entity, as a "thing." But if the "what" of an entity is specified by its substantial form, it is that form which determines whether the entity is animate or inanimate. Thus, in the case of an animal, that form is called an *anima* or *soul*, and in that of man, it is—for good reason—termed a *rational* soul. After death, in any case, when the soul has separated from the body, what remains of the body is no longer, strictly speaking, an entity, but what is termed a "mixture": a composite of inanimate substances, which as such is prone to decompose.

So much by way of recounting the most basic metaphysical conceptions. The notion we are about to introduce, on the other hand, pertains to what may perhaps be termed the deepest level of the Thomistic ontology. If it be the substantial form that founds a being (animate or inanimate), we now ask what it is that founds the substantial form itself: and that is what St. Thomas refers to as *the act of being*. As the Master himself explains: "The act-of-being is the most intimate element in anything, and the most profound element in all things, because it is like a form in regard to all that is in the thing."[1] The act-of-being proves thus to be none other than the cosmogenetic Act itself, viewed in relation to a specific being. And that act-of-being, moreover, bestows not only *being*, but also *a power to act with an efficacy of its own*. As Etienne Gilson observes magnificently:

> The universe, as represented by St. Thomas, is not a mass of inert bodies passively moved by a force which passes through them, but a collection of active beings each enjoying an efficacy delegated to it by God along with actual being.[2]

This brings us to a crucial point: on every level, that "efficacy delegated by God" entails a capacity, on the part of created beings, to achieve all manner of effects by way of *vertical* causation—beginning with the rudimentary examples pertaining to the inanimate realm previously cited in

1. *Summa Theologiae* I, 8, 1.
2. *The Christian Philosophy of St. Thomas Aquinas* (University of Notre Dame Press, 1994), 183.

the context of quantum theory. Turning now to the opposite end of the *scala naturae*, namely to the anthropic sphere, one sees that the prime example of vertical causation proves to be precisely what is traditionally termed "free will": for what else does the adjective "free" signify in this context than the capacity to act—not as a marionette, through the action of external causes, but indeed "from within." An act is "free," therefore, by virtue of the fact that it is not effected by a chain of external causes, but by the soul, the *anima* which acts from within, and hence by way of vertical causation.[3]

Moreover, in addition to "active" powers delegated by God, man is endowed with "cognitive" powers as well; and here too one finds that "verticality" in the sense of the "supra-temporal" proves to be key: for it turns out, as we shall see, that both visual and intellectual perception (the kind manifested, for instance, in understanding the proof of a mathematical theorem) demand a *supra-temporal* act. Now, to be "supra-temporal" means to be situated in the *nunc stans*: for between the seeming "now that moves" and the "now that stands" there is no middle term. Both of the aforesaid cognitive faculties presuppose therefore a capacity to "access" the *nunc stans*: to "step out of time," as it were. And that transcendence of time—of the temporal condition, if you will—has served from the start as the defining characteristic of the "vertical" in the sense we have made our own. In a word, the concept of "verticality" applies not only to causation, but to cognition as well.

---

3. The theoretical possibility of constraint "from above" by way of vertical causation has already been ruled out, and is in any case not normally under consideration when one speaks of "free will."

THREE VERTICAL POWERS OF THE SOUL

I will point out in passing that if every corporeal object gives rise to acts of vertical causality—as we have shown—it should come as no surprise that this capacity to activate VC increases as we ascend from the inorganic to the organic domain, and attains its zenith in man, the *anthropos*.[4] It should moreover be noted that between the extremes of the inorganic and the anthropic there must exist a plethora of "vertical" effects which a higher kind of science could presumably discover and put to good use, but which will remain undiscovered until it dawns upon the scientific community that causality exists actually in *two* modes: horizontal *and* vertical. When it comes to the ontologically higher domains, our sciences—as presently conceived—may actually be losing the better half of the picture. For those capable, on the other hand, of thinking "outside the box," the opportunities for the discovery of deeper and more powerful sciences appear to be vast.

FREE WILL AS VERTICAL CAUSALITY

THE CONNECTION BETWEEN "FREE WILL" AND VERTICAL causation came to light abruptly in 1998 when a mathematician named William Dembski published a remarkable theorem.[5] Having introduced the decisive concept of "complex specified information" or CSI, Dembski stunned the scientific world by proving—with complete mathematical rigor—that *no physical process, be it deterministic, random or stochastic*,[6] *can produce CSI.* But this means, in our termi-

4. It is to be understood that we are speaking of what may be termed the "sub-angelic" realm.
5. *The Design Inference* (Cambridge University Press, 1998).
6. A "stochastic" process is partly deterministic and partly random, as in the case of Brownian motion.

nology, that what has thus been disqualified from producing CSI is none other than *horizontal* causality! Wherever, therefore, we encounter the production of CSI, we have documented an act of *vertical* causation.

Now it appears that humans produce CSI almost incessantly; I myself, for instance, am apparently doing so at this very moment by writing these lines. And we take it in our stride: we see nothing remotely exceptional or astounding in these quotidian acts. Little do we realize that the metaphysical implications of this capacity are profound in the extreme, to the point that an inkling, even, of what actually transpires would in fact suffice to disqualify the contemporary Weltanschauung irrevocably: for the intervention of vertical causation actually entails a certain "access" on the part of man to the *nunc stans*, that elusive "now" that is said to "stand still," in which, according to Meister Eckhart, "*God creates the world and all things.*"

I should add that, in point of mathematical rigor, the aforesaid argument is yet incomplete: for inasmuch as we have access to an immense store of CSI—e.g., in the form of memory—it remains to show that the CSI we are said to "produce" (as I am doing right now, say, by composing this book) is not already "given" in that store. But whereas I fully acknowledge the logical force of this objection, I find it quite inconceivable that a chain of horizontal causation could take us—presumably by algorithmic means—from such "given CSI" to the "produced."[7] And I would add that by the time we recognize the verticality of visual perception and the non-algorithmic nature of mathematical proof—

---

7. Of course the question remains how the "given CSI" itself was produced. And here again Dembski's theorem implies that somewhere along the line an act of vertical causation *must have* intervened.

the subjects of the next two sections—there will be no more
need, in this context, for information-theoretic argument.
What I find remarkable is not that this kind of reasoning
can take us only so far, but rather that a science which
excludes all essence from its purview can in fact arrive at a
conclusion as powerful as Dembski's theorem.

THE VERTICALITY OF VISUAL PERCEPTION

IT HAS BEEN ASSUMED FOR CENTURIES THAT THE EYE IS IN
effect a camera, and that what we perceive is based upon a
retinal image. One is quite certain, moreover, not just that
this is the only reasonable way to account for visual perception, but that this "image theory" has been corroborated by
an abundance of evidence, beginning presumably with the
fact that spectacles affect vision in keeping with that paradigm. What we are not told, on the other hand, is that in the
40's of the last century, a cognitive psychologist at Cornell
University, named James Gibson, serendipitously uncovered
facts which *rigorously disqualify this visual-image theory*, and
that in the course of three subsequent decades of empirical
research he succeeded in replacing that now discredited paradigm by one which does square with the empirical facts.
Having dealt elsewhere[8] and at considerable length with this
epochal discovery, it will suffice to highlight the main findings which both reveal and establish the supra-temporal
nature of visual perception.

According to Gibson, what we perceive visually is not an
image—be it retinal, cerebral, or mental—but so-called
invariants given in what he terms *ambient*—as distinguished

8. *Science and Myth*, ch. 4.

from *radiant*—light. By "ambient" light he means light reflected from the environment, which Gibson conceives, not in "physical," but in what he refers to as "ecological" terms. To make a long story short: Gibson's use of the term "ecological" proves ultimately tantamount to what I term *corporeal*, the point being that Gibson's "environment" owns not only quantitative attributes, but *qualities* as well. What then are the "invariants" we are said to "pick up" in the act of visual perception? Gibson, to be sure, speaks of these in his own "research-based" terms; yet in light of traditional metaphysics, what actually stands at issue are none other than *forms*. And let us note at once: if what we "pick up" in the act of visual perception are indeed forms, then—and then only!—is it possible to perceive, not simply an image or phantasm, but the very objects: that is to say, *the external world itself*. For in that case we perceive—not a mere image or *effect* of that world—but the very forms that constitute its reality.

Gibson's discovery, as noted before, amounts thus to *a scientific refutation of the Cartesian doctrine*: specifically its epistemology, which affirms that the object of perception constitutes a mere phantasm or "thing of the mind." Even as Heisenberg's physics has demonstrated that there exists actually no such thing as a Cartesian *res extensa*, so has Gibson's discovery toppled the second pillar of the Cartesian edifice: the misbegotten notion of *res cogitans*. The amazing fact—which Gibson himself clearly recognized—is that his so-called "ecological" theory of visual perception has validated the immemorial premise that we do actually "look out" upon the world, which thus proves to be—not the world as conceived or imagined by the physicist—but none other than what we have termed the *corporeal*.

It is essential to recognize that this radical shift in our

understanding of *what* we actually perceive necessitates a corresponding change in our conception of how that perception is effected. What needs above all to be understood is that so long as the faculties of perception are themselves a mere aggregate or "sum of parts," a pick-up of *forms* is simply inconceivable. So long, therefore, as one conceives of the percipient in what might be referred to as "post-Enlightenment" terms, an understanding of perception proves thus to be impossible: that very premise suffices to render it such. And amazingly, as time went on, Gibson the quintessential empiricist came to realize this ontological fact with ever greater clarity: his experimental findings actually demand nothing less. It turns out that one cannot in fact understand visual perception simply by looking at the retina or at the behavior of neurons in affected parts of the brain. This is not to say, of course, that these processes have nothing to do with visual perception; the point, rather, is that *they are part of a process which in truth is more than the sum of its parts.* And it came to pass that Gibson—this genius of an empiricist—came, by gradual and meticulous steps, to understand that the mystery of perception resides precisely in this "more." It remains only to point out that this "more" is finally none other than what is traditionally termed a "soul."

The power of visual perception derives thus, in the final count, from the *soul.* And inasmuch as the soul is united to the body as its substantial form, what to the neuro-scientist appears as a vast ensemble of neuronal activity *constitutes in reality a single event, a single act of the living and sentient organism.* The point is simple: if perception were merely a matter of neuronal firings, we would need a homunculus—a little man within the brain—to "read" these events—which is of course absurd. It should therefore come as no surprise when Sir Francis Crick (of DNA fame) informs us that "*we can see how the brain takes the picture apart, but we do not yet see how it puts it together*"[9]—the point being that the brain itself *cannot* in fact "put it together" at all: it takes the *soul* to accomplish that prodigy. And the reason, moreover, why it *can* do so resides in the ontological fact that *the soul is not subject to the bounds of space.* Let us note what this entails: it is by virtue of the resultant "ubiquity" that the soul can actually be present to each cell in the body—not as some minute fragment of itself—but in its undivided and undiminished entirety.[10]

Yet it turns out that even this marvelous capacity—this seemingly miraculous ability on the part of the soul to transcend spatial bounds—does not yet suffice: for it happens that the prodigy of visual perception requires a comparable transcendence of *temporal bounds* as well! And this too came to light in the course of Gibson's experiments, culminating in the recognition that we perceive motion, not "moment by moment," but the only way motion *can* in fact be perceived: *all at once!* Yet, amazing as this fact may be, it is hardly surprising: for if in fact we did perceive "moment by moment," we would in principle fare no better than a camera, which is able to "see" merely a succession of images while the motion itself remains invisible.

I would emphasize that Gibson's arguments are both subtle and fully rigorous once the point has been grasped, and confirm unequivocally that visual perception constitutes indeed a *vertical* act. Plato was right: so far from being con-

9. *The Astonishing Hypothesis* (New York: Simon & Schuster, 1995), 159.

10. Once one catches so much as a glimpse of what the transcendence of space—the existence of the intermediary domain—actually entails, one's outlook regarding contemporary science will never be the same.

fined to the temporal realm, the soul does in truth *have access to eternity.*

THE VERTICALITY OF INTELLECT

ONCE IT HAD BEEN ASSUMED THAT THE EXTERNAL WORLD of *res extensae* constitutes a single gigantic mechanism, it did not take long for the enlightened ones to draw the conclusion that man too is basically a mechanism. And this entails that his so-called intellectual capacities must likewise be the result of some—presumably cerebral—machinery. Around the fateful year 1900, moreover, the basic component of that hypothesized mechanism—the so-called neuron—was identified by a Spanish biologist named Ramón y Cajal; and that discovery, as one might surmise, gave rise to a world-wide brain-research enterprise designed to ascertain the structure and *modus operandi* of that horrendously complex "machine" made up of neurons: close to 100 billion in all.

Around the year 1936, moreover, another fundamental breakthrough occurred, which soon proved to be related. It happened when a mathematician, named Alan Turing, posed the question whether mathematical reasoning as such could, in principle, be carried out by a "machine" or mechanism of some sort. Now the kind of "reasoning" Turing had in mind has since been termed *algorithmic*; and what he did—to the amazement of the mathematical world—was to construct what has ever since been termed a Turing machine: a device, existing "on paper" if you will, capable in principle of executing every conceivable algorithm by means of a corresponding "program" inscribed in the "software" part of the machine. The fact is that Turing discovered the universal prototype of which every computer in the world constitutes a partial embodiment.

In the wake of this epochal discovery it came to be widely assumed that the human brain functions—at least in part— as a computer, and that human intelligence is consequently *algorithmic*. There can be little doubt, moreover, that this has been the reigning paradigm in the pertinent fields of scientific endeavor ever since the discovery of the Turing machine: that to this day, in fact, it constitutes the underlying premise upon which our scientific understanding of "human intelligence" is based. It may therefore come as a shock to many that this putatively "scientific" tenet turns out to be not only questionable, but *has in fact been disproved with mathematical rigor*. That is what I wish now to explain as briefly as I can.

The argument is based upon a stupendous theorem, proved in 1933 by a 25-year-old mathematician named Kurt Gödel, which affirms that given a set of mathematical axioms (extensive enough to encompass arithmetical propositions), there must exist a proposition which is *true* but *unprovable* in that system. The proof of Gödel's theorem is based upon the amazing fact that not only arithmetical propositions, but all possible proofs of such within a given axiom system can be "ordered": that is to say, indexed by the natural numbers 1, 2, 3... Now this is the part of the proof which is both technical and difficult in the extreme, whereas what remains—though highly ingenious—turns out to be rather simple. Here then is that second part, which concludes Gödel's argument:

By virtue of Part I, we may assume that there exists a function $P(m,n)$, defined for all natural numbers m and n, such that, for every m, $P(m,n)$ is a propositional function of n (an algebraic statement depending on n, which may be true or false) and a function $\Pi(k)$ which orders all mathematical proofs in the given axiom system.

We now define the following propositional function: "There exists no k such that $\Pi(k)$ proves $P(w,w)$." Since our enumeration $P(m,n)$ of arithmetical propositions is complete, there must exist a natural number s such that $P(s,n)$ is the aforesaid function. Now consider the proposition $P(s,s)$: the first thing to note is that this proposition is unprovable (since our construction entails that "there exists no k such that $\Pi(k)$ proves $P(s,s)$");[11] and the second is that $P(s,s)$ is true: for indeed there exists no k such that $\Pi(k)$ proves $P(s,s)$. *And this is how Gödel proved the existence of true but unprovable propositions!*

The question, now, is whether this theorem has been proved by way of an algorithm or not. It is always possible, of course, to claim that an algorithm of some kind has unwittingly played out in the process of arriving at a given conclusion; but as Roger Penrose observes, the algorithms used in mathematics are well known and communicable, and it is evident in the case of Gödel's theorem that no such has come into play. In a word, the intelligence which enabled us to understand the proof in question is manifestly non-algorithmic. "When we convince ourselves of the validity of Gödel's theorem," Penrose goes on to say, "we not only 'see' it, but in so doing we reveal the very non-algorithmic nature of the 'seeing' process itself."[12] Now, we concur with this statement wholeheartedly—except for one word: we must insist that this "seeing" is not in fact a "process," but a *vertical* and therefore *instantaneous* act. As the expression goes, it is manifestly a question of "seeing the point"—and a "point" is to be seen "all at once" or not at all.

11. The reader will note that the propositional functions $P(k,k)$ and $P(s,s)$ are evidently one and the same.
12. *The Emperor's New Mind* (Oxford University Press, 1989), 418.

One should add that precisely the same actually holds true when a theorem is supposedly proved by an algorithm: for it is actually not the algorithm that proves the theorem, but the person who "sees" that it does. Strictly speaking, a formal proof of a mathematical theorem can do no more than elicit, in those who are qualified, a perception of its validity. The fact that an argument or chain of reasoning constitutes a proof by virtue of meeting appropriate criteria of validity is of course undeniable—but that does not obviate the necessity of "seeing the point." Thus, in the final count, science is indeed *"nothing but perception,"* as Plato noted long ago. In the end—when the work has been accomplished and the quest attained its goal—that's what it reduces to. And that consummation, let us add, is achieved—not by the rational faculty, which is discursive and operates in time—but by the *intellect*, properly so called, which does *not* operate in time, but in what has been termed the *nunc stans*. For the intellect is indeed the "eye of the soul" by which we *see*: the "part," as Plato says, which "pertains to eternity."

5

THE WAR ON DESIGN

SINCE THE SEVENTEENTH CENTURY, WESTERN civilization has been subject to the spell of a new and supposedly "scientific" Weltanschauung, initiated by Galileo, Descartes and Newton. From the publication of Newton's *Principia* in 1687 to the discovery of quantum physics in the early twentieth century, it was assumed by the pundits of the Enlightenment that, at bottom, the universe constitutes a gigantic "clockwork," in which the disposition of the parts determines—with mathematical precision!— the movement of the whole. And even in the face of the quantum facts, that paradigm was not actually discarded, but merely modified. To this day one is prone to conceive of the universe basically as a "clockwork," howbeit one which no longer functions with absolute precision: one could say that in addition to rigid cogwheels, it now comprises some "wobbly" components as well, which in effect play the role of "dice." The larger picture, thus, has scarcely changed at all: now as before, Nature is perceived, on scientific authority, as constituting precisely what Whitehead referred to as "a dull affair": merely "the hurrying of material, endlessly, meaninglessly."[1]

With the appearance of what we have termed "vertical causation," however, the picture has radically changed. I

---

1. *Science and the Modern World* (New York: Macmillan, 1953), 54.

need hardly recount how, beginning with the quantum measurement problem, vertical causation has come into view time and again, profoundly affecting the entire spectrum of scientific domains—from physics to cognitive psychology—touching even, if you will, upon the mathematical sciences.

In light of these findings—and this is the first major point I wish now to convey—it emerges that our natural sciences have presently reached a stage at which the next round of foundational discovery may very well hinge upon a recognition of vertical causality. I say "foundational," because it is of course always possible to investigate this or that phenomenon along already established lines, and in so doing reap whatever technological or other benefits may ensue. Obviously, however, I am referring to something quite different: to the question, for instance, how the mysterious and hitherto inexplicable "epigenome" fits into genetics as presently conceived, or what neuronal structures can and cannot accomplish by way of facilitating thought, perception, memory and the like.[2] If not perhaps the first, then most assuredly the second question requires that we bring the concept of vertical causation into play. I surmise that when it comes to research of a truly foundational kind, we are approaching the end of what can in principle be understood on the basis of horizontal causality alone, and that much of what presently impedes progress at the frontiers of scientific inquiry may prove, ultimately, to be an effect of VC. What is called for, I maintain, is a vastly deepened understanding of Nature, based upon the recognition that horizontal causality, so far from standing alone, is perforce complemented in

2. I have dealt with this question in a general way in "Neurons and Mind." See *Science and Myth*.

the final count by vertical modes of causation. It is high time, I would argue, to jettison our Galilean, Cartesian, and Newtonian assumptions and become philosophically literate once again.

✦ ✦ ✦

INASMUCH AS VERTICAL CAUSATION PRESENTS ITSELF IN numerous modes, some fundamental distinctions need now to be made. The primary dichotomy stems from the fundamental recognition touched upon earlier: the fact, namely, that *the act of being* bestows upon creatures not only existence by way of a substantial form, but also an efficacy, a certain power to act. To be precise, every corporeal entity is endowed with a capacity to act by way of vertical causation; and inasmuch as this power pertains to the substantial form, we may refer to it as *substantial* VC. It is also needful, however, to recognize a higher mode of vertical causality, which does not emanate from a substantial form, but can in fact *give rise to* substantial forms, a mode which we may consequently refer to as *creative*. And this brings us finally to the subject of the present chapter: the notion of *design*. What do we mean by this term? One is prone to say that "design" is simply an effect of creative VC; but this is far too broad to be of much use: for what then, in the universe, would *not* be an instance of design? To be of interest, the concept should be restricted to what in fact *cannot* be produced by "the hurrying of material, endlessly, meaninglessly." There are then, first of all, two instances of design which prove to be of special interest, one terrestrial and the other cosmic: *speciation* namely, and a marvelous order pertaining to the cosmos at large, which we term *immobility*, a concept that will emerge in the course of our inquiry.

Which brings me finally to the subject of the present chapter: "The War on Design." My first contention should hardly come as a surprise: I charge that the Darwinist claim, so far from constituting a scientific hypothesis supported by empirical evidence, proves to be in truth an *ideological* tenet, based upon the *a priori* denial of design in the form of speciation.[3] It is my second claim, rather, that may cause astonishment: I shall contend that even as Darwinism rests upon the denial of design in the origin of species, relativistic physics at large is based upon the *a priori* rejection of design in the form of *immobility*. Einsteinian physics proves thus to be a kind of Darwinism on a cosmic scale; and turns out in the end—to the surprise and consternation of many—to be likewise untenable.

THE THEORETICAL
BASIS OF EINSTEINIAN PHYSICS

LET US BEGIN BY REFLECTING UPON EINSTEIN'S 1905 PAPER, which inaugurates his so-called "special" theory of relativity.[4] The ostensible mission of that article—its accomplishment one might say—was to impose a condition upon physics referred to as the Principle of Relativity. But what is it, precisely, that motivated Albert Einstein to modify the classical equations, to render them "relativistic"? Why *should*

---

3. It is safe to say that with the discovery of DNA around the middle of the last century Darwinism was in effect disqualified as a scientific theory. With the publication moreover of Dembski's 1998 theorem regarding "complex specified information" it has been rigorously disproved on mathematical grounds, and thus reduced from a bona-fide scientific hypothesis to the status of a sociological phenomenon.

4. As distinguished from his "general" theory, published in 1917.

the laws of physics be "invariant" in that specific sense? Let us see how Einstein himself responds to this question.

In the survey of his theory, special plus general, published under the title *The Meaning of Relativity*,[5] Einstein begins by recalling the basic facts of analytic geometry, starting with the concept of a Cartesian coordinate system. For the sake of concreteness, let's suppose for the moment that we are dealing with the Euclidean plane in which a point O (termed "the origin") has been specified. A Cartesian coordinate system consists then of two mutually perpendicular and oriented lines through O. With every point P in the plane one can now associate a pair of coordinates (x_1, x_2), defined by the perpendicular projections of P onto the corresponding axes.[6] It follows—by the venerable Pythagorean theorem[7] no less—that the square of the distance OP equals $x_1^2 + x_2^2$. What is crucial here is the fact that this relation—this "law"—holds in *every* Cartesian coordinate system, that it is thus "*invariant*."

Now, simple and elementary as these considerations may be, they can serve admirably as a point of departure for an introduction to Einsteinian physics. Getting back to our example: let us note that a coordinate system effects a transition from a *geometric* to an *analytical* structure, which obviously depends upon the coordinate system we happen to choose. The problem, then, is to discover what, in a given representation, is independent of that choice: is thus "geometrical" like the expression $x_1^2 + x_2^2$. It is needful, in other

5. Princeton University Press, 1955.

6. To be precise: If Q_i denotes the projection of P onto the ith axis, we define x_i to be the distance (positive or negative) from O to Q_i.

7. "The sum of the squares of the sides of a right triangle equals the square of the hypotenuse."

words, to distinguish the "non-geometrical" elements introduced by this construction from properties indigenous to the Euclidean space: and this is where the all-important notion of *invariance* comes into play. In the case of Cartesian coordinates in the Euclidean plane, the term $x_1^2 + x_2^2$ is a prime example of an invariant: as we have noted, it equals the length squared of the line segment OP.

The key conception of Einsteinian physics can now be explained: for it happens that Albert Einstein was fascinated by geometry, and conceived of physics in basically geometric terms. What now takes the place of the Euclidean plane is the locus of all "points in space" and "moments in time," which has since come to be known as the *space-time continuum*. Einstein's objective—his momentous idea—was to reduce physics in effect to a geometry on this four-dimensional manifold. But of course, Einstein the quintessential physicist conceived of that space-time, not as a mathematical abstraction, but—true to Lord Kelvin's definition of physics as "the science of measurement"—as based upon mensuration.

What then constitutes an Einsteinian "coordinate system"? The question reduces to this: Given a point P in space-time, how can one associate an ordered set of four real numbers (x_1, x_2, x_3, x_4) with that point? One can do so, in principle, by means of a reference *frame* K, which we may picture as consisting of three mutually perpendicular rods emanating from a point O. It is apparent that such a frame enables us "in principle" to assign four coordinates to every "point" P in space-time: three spatial coordinates (x_1, x_2, x_3) namely, plus a "time-coordinate" t as measured by a clock stationary with respect to K.

The next step in the construction of Einsteinian "space-time" geometry is to define the analog of "distance": i.e., to specify what plays the role of $x_1^2 + x_2^2$ in the Euclidean plane. And for good mathematical reasons, Einstein chose the quadratic form $x_1^2 + x_2^2 + x_3^2 - c^2 t^2$, where c denotes the speed of light. In the first place, since it obviously makes no sense to add, say, *seconds* to *meters*, it is necessary to multiply t by a *velocity* to obtain a *distance*. But why choose the light velocity c? And again, for good reasons: apart from the fact that there are grounds to suppose that c constitutes a universal constant, it happens that coordinate transformations which preserve the given quadratic form—the so-called Lorentz transformations—preserve the Maxwell equations of electromagnetism as well. By this choice Einstein has therefore taken the decisive step in the construction of a physics in which the phenomena of electromagnetism can be viewed in geometric terms. One thing more, however, needs to be done, and that is to specify the actual reference frames—the so-called *inertial* frames—which in this space-time geometry will play the role of a Cartesian coordinate system: then only does the Einsteinian space-time acquire a physical sense, and thus engender predictions which can actually be put to the test.

✦ ✦ ✦

HOW THEN DOES EINSTEIN DEFINE THE CLASS OF *INERTIAL frames* K? He does so in his 1905 debut by what he terms "*the principle of special relativity*,"[8] which asserts that *if K is inertial and K' moves uniformly and without rotation with respect to K, then K' is likewise inertial.* One consequently needs but to identify a single inertial frame K_0 to determine the entire class.

8. The adjective "special" does not appear in the 1905 paper, but was added after the so-called "general theory of relativity" appeared.

Let us suppose now that we have identified such a "first" reference frame K_0, and that the classical (i.e., pre-relativistic and pre-quantum) equations of physics hold in K_0. The question, now, is whether they hold likewise in every other inertial reference frame K. If so, then classical physics satisfies Einstein's principle of special relativity; and of course: if not, it doesn't. To resolve this issue we need first of all to recall that the classical equations of physics divide into two groups: the equations of mechanics, which go back to Sir Isaac Newton's *Principia*, published in 1687, and the equations of electromagnetism, formulated 178 years later by Clerk Maxwell. Now we know from the start that the equations of electromagnetism do satisfy Einstein's principle of special relativity, by the fact that they are Lorentz invariant. As we have pointed out, the Einsteinian geometry is actually "tailor-made" to render the Maxwell equations "geometrical"! The real test of Einsteinian relativity, therefore, comes from the side of mechanics: do these classical Newtonian equations satisfy Einstein's Principle or do they not? And it is obvious, by the fact that these equations are *not* Lorentz invariant, that indeed they don't.

To summarize up to this point: Albert Einstein has constructed a "relativistic" physics, conceived in terms of a space-time geometry, which is compatible with the Maxwell theory of electromagnetism but *not* with Newtonian mechanics. Now this leaves him, obviously, with two options: to reject his so-called "special theory of relativity" on the grounds that it does not square with the equations of mechanics, or to alter these equations—to render them "relativistic" by fiat as it were—to save his theory. And needless to say, Einstein chose the second course: the Procrustean option, his critics might say.

THE WAR ON DESIGN

✦ ✦ ✦

TWO POSSIBILITIES REMAIN: EITHER EINSTEIN IS RIGHT and the equations of classical mechanics need indeed to be revised, or the equations of classical mechanics are correct as they stand, and it is actually his "relativistic" mechanics that proves to be false. How, then, does Einstein argue his case: on what grounds does he justify his theory? To answer this question we turn now to his original 1905 paper, published in the *Annalen der Physik*—under the rather unassuming title "*On the electrodynamics of moving bodies*"—which marks the birth of "relativistic" physics, to see how Einstein himself justified his revolutionary proposal.

As might be expected, he begins by recounting experiments pertaining to the domain of *electromagnetism* to exhibit observable phenomena which do "depend only on relative motion," that is to say, satisfy his stipulated Principle of Relativity.[9] But what about the "other half" of classical physics: the equations of mechanics? Einstein gives a very brief (and interesting!) answer to this question. Having alluded—in the opening paragraph of his paper—to certain electromagnetic phenomena (which, as we have noted, comply with his Principle), he begins the second paragraph as follows: "Examples of this sort," he writes, "together with the unsuccessful attempts to discover any motion of the Earth relative to the 'light medium,' suggest that the phenomena of electrodynamics *as well as of mechanics* [my italics] possess no properties corresponding to the idea of absolute rest. They suggest rather that, as has already been shown to the first order of small quantities, the same laws of electrodynamics and optics will be valid for all frames of ref-

9. The adjective "special" was introduced later, as noted above.

erence for which the equations of mechanics hold good. We will raise this conjecture (hereafter referred to as the 'Principle of Relativity') to the status of a postulate."

The reasoning here is astonishing. Let us first of all consider Einstein's allusion to the Michelson-Morley experiment of 1887, which, as one knows, was designed to detect and measure the postulated orbital velocity of the Earth around the Sun (said to be around 30 km/sec), but proved "unsuccessful" inasmuch as no such velocity was found. On what basis therefore, let us ask, does Einstein rule out the theoretical possibility that the experiment may actually have proved that in fact *there is no* such "orbital" velocity, as the Michelson-Morley finding appears to attest? After all, given that the question pertains to physical science *on its most fundamental level*, it would seem that no conceptual possibility—no matter how apparently improbable—should be ruled out simply "by a wave of the hand"! And as a matter of fact: to imply that a "uniform motion without rotation" is not detectable is to beg the very question which stands at issue: namely, whether or not the Einsteinian Principle of Relativity is true.

What I find still more surprising, however, is Einstein's claim that empirical findings "suggest that the phenomena of electrodynamics *as well as of mechanics*" conform to his postulate: for whereas, in the case of electrodynamics, the property in question follows mathematically from the fact that Maxwell's equations for the electromagnetic field are in fact Lorentz invariant, the opposite holds true for the classical equations of mechanics. For unlike electrodynamics, classical mechanics happens *not* to be Lorentz invariant; and whereas it was consequently a foregone conclusion that "the phenomena of electrodynamics" conform to Einstein's Principle, the opposite holds true in the case of mechanics.

In other words, even as the phenomena of electrodynamics *predictably* obey the Principle of Relativity, so on the strength of classical physics "the phenomena of mechanics" *predictably* do not. Now to be sure, Einstein was well aware of the fact that the classical equations of mechanics do *not* obey his Principle—which is after all the reason he felt obliged to alter them, to render them "relativistic." My point is that this fateful step was, in the final count, authorized by nothing more substantial than the failure of the Michelson-Morley experiment to detect the postulated—but never yet observed!—"orbital" velocity of the Earth.

There is thus no complete—let alone compelling—argument to justify the shift from classical to relativistic physics. Einstein's pivotal allusion to the Michelson-Morley experiment has no doubt a powerful *psychological* efficacy within the scientific community, but actually carries no weight: given that the experiment was designed to detect and measure—for the very first time—a conjectured velocity, its failure to confirm that hypothetical motion hardly disproves the validity of classical mechanics!

Moreover, there is nothing in any way illogical, incongruous, or scientifically objectionable in the fact that the equations governing mechanical phenomena and those which describe electromagnetic fields should be invariant under different groups (Galilean and Lorentz, respectively). It could in fact be argued quite cogently that in view of the radically different nature of these respective domains, such a discrepancy is rather to be expected. In any case, there appears to be no bona-fide argument which would exclude that possibility, and Einstein's above-cited reference to "the phenomena of mechanics" most certainly doesn't alter this fact.

Meanwhile it is a mathematical fact that, so far from satisfying Einstein's Principle of Relativity, the equations of classical physics—mechanics plus electromagnetism—actually entail the very opposite: they imply, namely, that if the equations of mechanics *and* of electromagnetism both hold in two reference frames K_0 and K, then K is in fact *stationary* with respect to K_0. Instead of the Einsteinian Principle of Relativity, we actually have here a Principle of *Immobility*— and not as a conjecture in conflict with the equations of mechanics, but as a *theorem of classical physics*. The fact is that pre-Einsteinian physics implies the very opposite of Einstein's postulate: in place of an ensemble of "inertial" reference frames in uniform motion with respect to each other, in which none can be singled out on physical grounds, one finds that *physics itself defines a state of absolute rest.*

So far as I know, neither Albert Einstein nor any post-Einsteinian physicist of repute has ever so much as mentioned this remarkable fact, let alone explored its implications. No one among the *avant-garde* physicists appears to have seriously entertained the possibility that *the equations of classical mechanics may actually be correct!* In place of a Principle of Immobility corroborated empirically for well over half a century,[10] they have opted for a Principle of Relativity for which there is actually no empirical validation at all: that *conjectured* Principle takes precedence, in their mind, over three centuries of scientific verification![11] But why? On what grounds does Einstein justify his rejection of classical mechanics? When and where, exactly, did it fail? Does he

10. Since 1865 in fact, when Clerk Maxwell wrote down the equations of the electromagnetic field.

11. The equations of classical mechanics, which Einstein rejected, have been in use since 1687.

have nothing more cogent to offer in that regard than the reputed "failure" of the Michelson-Morley experiment?

I contend that the Einsteinian preference for the Principle of Relativity is based in the final count—not on scientific or empirical grounds—but on *ideological* premises. The fact is that the existence of *immobile* reference frames is closely tied to geocentrism, which constitutes—unmistakably!—an instance of *design*. Considering then that "design" implicates a Designer, it is hardly surprising that geocentrism has long been taboo to the scientific elite; for as Richard Lewontin, speaking for the scientific community at large, apprises us: "*We cannot allow a Divine Foot in the Door.*"

✦ ✦ ✦

CLASSICAL PHYSICS, BY VIRTUE OF WHAT WE HAVE TERMED the Principle of Immobility, affirms the existence of "*stationary*" reference frames: frames K, namely, in which the classical equations of mechanics and of electromagnetism both hold. And this brings us to the crucial question: what about *geocentric* reference frames (frames at rest with respect to the Earth): are these in fact stationary, or are they not? If they are, this would mean that *classical physics entails geocentrism*!

What evidently stands in the way of this conclusion is the seemingly sacrosanct tenet that the Earth rotates around its polar axis: every 24 hours, to be exact. One needs therefore to ask oneself whether this reputed fact has been rigorously established by scientific means. Of course, if the question were asked, the overwhelming majority of respondents— from laymen to astrophysicists—would, without a moment's hesitation, answer in the affirmative. Yet it happens that this virtually unanimous verdict proves to be incorrect:

for it follows, by what is termed *Mach's Principle*, that it is not in fact possible to ascertain by empirical means whether it is the Earth that rotates while the cosmos at large stands still, or whether, on the contrary, it is the cosmos at large that rotates diurnally around a stationary Earth.

This is all we need to know: it permits us to affirm without fear of contradiction, based squarely upon classical physics, that *geocentric reference frames are stationary*. And thus one arrives at what is in effect the pre-Copernican cosmography, which conceives the Earth as an immobile sphere situated at the center of the universe, which consequently revolves diurnally around the Earth's polar axis. It appears that Einstein's 1905 Principle of Relativity, so far from being necessitated on scientific grounds, was based on an *a priori* rejection of what I have termed the Principle of Immobility, and thus of classical physics *per se*, which entails that Principle. The modifications imposed upon the equations of mechanics to render them Lorentz invariant are therefore unjustified on physical grounds: there *are no* empirical findings which necessitate the alterations Einstein imposed. And as regards our use of Mach's Principle (or half of it, to be exact), it may not be without interest to note that Einstein himself was profoundly inspired by this discovery: it is he, in fact, who coined its name. It strikes me as ironic, thus, that the very Principle which may have set him on the spoor of his "general" theory actually validates what he seemed to abhor: geocentrism no less!

But lest there remain any doubt regarding the validity of taking geocentric reference frames to be stationary, let me refer to a recent paper—entitled *"Newton-Machian analysis of a Neo-tychonian model of planetary motions"*[12]—in which

12. *Eur. J. Phys.* 34, 2 (2013): 383.

a physicist named Luka Popov calculates planetary orbits by means of Newtonian physics, based on a *geocentric* reference frame. In so doing, Popov has abrogated, once and for all, the hegemony of heliocentrism, inaugurated in 1687 by Isaac Newton himself through the quasi-heliocentric[13] calculation of planetary orbits, based upon his newly-discovered "laws of motion" and infinitesimal calculus. Ever since that monumental breakthrough it has been an unquestioned dogma of Western science that "planet Earth" revolves around the Sun: once a year to be exact, with an orbital velocity on the order of 30 km/sec as we have been often enough reminded. But whereas the validity and indeed the accuracy of the Newtonian calculations have never been in doubt, Luka Popov appears to be the first physicist in history *to disprove* that they entail the aforesaid dogma: the notion, namely, that the Earth is a planet revolving around the Sun, or even permit us to infer that the Sun itself *is not* a planet revolving around the Earth. He rather confirms, by direct "Newton-Machian" calculation, what classical physics, in conjunction with Mach's Principle, permits us to conclude: the fact, namely, that the "heliocentric" cosmography can actually be obtained via Newtonian physics on a *geocentric* basis.[14]

Let me repeat: *Luka Popov's result has broken the long-standing hegemony of heliocentrism.* It demonstrates, on the

---

13. I say "quasi-heliocentric" because Newton placed the origin of his coordinate system, not at the center of the Sun, but at the center of gravity of the Sun-planet 2-body system.

14. The term "Neo-tychonian" need not detain us. Tycho Brahe, in the 16th century, conceived the remarkable idea that whereas the Sun rotates around the Earth, the remaining planets rotate around the Sun. This formulation embodies all the advantages of the Copernican theory, to which it is in fact isomorphic.

basis of Newtonian mechanics, that it is equally legitimate to claim that the Sun rotates (annually) around the Earth: it all depends on your choice of coordinates, *and geocentric coordinates are in fact legitimate.*

The picture changes drastically, however, the moment one takes electromagnetism into account: for now the Principle of Immobility comes into play, which, as we have seen, singles out the Earth from all other celestial bodies by the fact that it can be both stationary and central. What confronts us here constitutes evidently the very pinnacle of design: no wonder "relativists" of every stripe abhor the notion like the plague! So far from being a planet, the Earth can thus be viewed as the very antithesis: as the stationary center, namely, around which all other celestial bodies are constrained to revolve diurnally. And as to the authentic planets or "wanderers," beginning with the Sun: these do then execute, in addition, their appointed orbits around the Earth, very much as the ancient astronomers had ascertained.

✦ ✦ ✦

GETTING BACK TO ALBERT EINSTEIN: DESPITE OCCASIONAL references to "the Old One" (*"der Alte"*), he is evidently opposed to the notion that "a Divine Foot" could have a detectable effect upon the physical universe. He appears rather to accept the premise that "horizontal" causality by itself suffices to explain all that could possibly be of interest to the physicist, and moreover made it his task to discover the "ultimate" differential equations: the ones which supposedly account, with one hundred percent accuracy, for all that is measurable. He was irrevocably opposed, therefore, to the notion of quantum indeterminacy, convinced apparently that "beneath" the quantum level, horizontal causality reigns supreme. And based presumably upon these certainties, the great physicist pursued the one and only course left open: which is *to alter the classical equations of mechanics—* by fiat as it were—*to render them "relativistic."*

What I wish to emphasize is that Einsteinian relativity is actually predicated upon the assumption that *there can be no such thing as an immobile reference frame, a K_0 "at rest"*: it is this denial that leads quite naturally to at least the special theory of relativity. But given that there exists not a shred of empirical evidence in support of that denial, one sees that Einsteinian physics cannot but be based ultimately on ideological grounds. Yes, we *are* witnessing a "War Against Design": it is by no means a pious fiction! Modern science *is* not—and never *has been*—the "disinterested quest of truth" our textbook wisdom proclaims it to be. The example, to be sure, which most flagrantly contradicts that official narrative—to the point that any "unprogrammed" spectator can readily see it for himself—is doubtless the Darwinist theory of evolution, the hegemony of which remains intact, as we have seen, even after that hypothesis has been *mathematically* disproved.[15] The case of Einsteinian physics is of course far more difficult to "unmask"; and yet, the moment one brings "geocentrism" into the picture, the pieces fall quite readily into place. The first two paragraphs of Einstein's 1905 paper suffice already to raise the definitive questions: all one need do, when Einstein postulates his Principle of Relativity, is ask "*why?*": why alter the Newtonian equations? why suppose that they are in any way deficient? what experiment has told us so? Is it the Michelson-Morley? But in that case, why fault the classical equations of mechanics for the failure to detect a motion which *no previ-*

15. See footnote 3 above.

ous experiment has ever verified? Was the Michelson-Morley experiment not in fact designed precisely to verify that hypothesis? Is it sound scientific practice, then, to alter the fundamental equations of physics when they fail to accord with a preconceived conjecture?

The decisive fact is that the Einsteinian postulate is actually based not on scientific but on *philosophical* grounds. What stands at issue is not in truth a question of physics, but of philosophy: a *philosophy of physics* if you will. I see no need, moreover, to delineate that philosophical position, to spell out its principles and tenets. Suffice it to say that these evidently preclude the cosmos from being endowed with a structure which cannot, in principle, be accounted for in terms of *horizontal* causation alone. And that accounts for the Einsteinian animus against geocentrism: for, like it or not, geocentrism calls for a *vertical* mode of causality.

THE EMPIRICAL ARGUMENT
CONTRA EINSTEINIAN PHYSICS

GIVEN THAT THE THEORETICAL FOUNDATIONS OF RELATIV-istic physics prove to be inherently ideological, it behooves us to consider the empirical side of the question: to examine how the Einsteinian theory fares when put to the test. And it will actually suffice to consider the special theory: for if this proves to be untenable, so does the general theory as well.

Let us recall in the first place that the special theory of relativity differs from classical physics only in the equations of mechanics, which it has altered by inserting the square root of $(1-v^2/c^2)$ in certain locations, v being an observable velocity and c the speed of light. Now, given the enormous magnitude of c (approximately 300,000 km/sec) as compared to a normally observable velocity v, one sees that the factor in question will tend to be so close to 1 as to render its effect unmeasurable. If v = 1000 km/hr, for example, the square root of $(1 - v^2/c^2)$ will be 0.99999945...! The empirical verification of Einsteinian physics proves thus to be "challenging," to say the least.

It behooves us, first of all, to consider the fateful formula $E = mc^2$, which almost everyone in the world attributes to Albert Einstein's theory. Despite the fact, however, that Einstein did derive this formula from his special theory of relativity, it stems actually from its classical part: i.e., from the Maxwell equations for electromagnetic fields, which go back to 1865. The famous formula has consequently no bearing whatsoever on relativistic physics, a fact Einstein himself admitted in 1950.[16] Obviously, however, in the interim that fateful formula came to be viewed worldwide as the consummate vindication of Einstein's theory: what indeed could be more convincing than the explosion of an atom bomb?

Misconceptions aside, the question whether special relativity has passed empirical muster proves to be an incurably technical issue. And no wonder, if at a speed of 1000 km/hr one needs to differentiate between 1 and 0.99999945! In his ground-breaking study,[17] Robert Bennett surveys more than three dozen experiments, covering a broad spectrum of empirical domains, which unfortunately hardly lend them-

16. *Out of My Later Years* (New York: Philosophical Library, 1950), 282. For a brief history of the "atom bomb equation" I refer to Appendix 2 in the encyclopedic volume by Robert A. Sungenis and Robert J. Bennett, entitled *Galileo Was Wrong, the Church Was Right* (Catholic Apologetics International Publishing, 2008). Chapter 10 provides a concise and magisterial overview of the empirical literature bearing upon Einsteinian physics.
17. Ibid.

selves to summary exposition comprehensible to non-specialists.

There are exceptions. For example, Einstein himself mentions the so-called Fizeau relation for the propagation of light in moving media as a confirmation of special relativity.[18] What stands at issue here is the speed of light in a moving liquid; and as might be expected, the subject proves indeed to be technical, involving such things as "the Fresnel drag coefficient" and the physics of Fizeau's "optical interferometer." But happily, one need not have the slightest idea what Fresnel and Fizeau were talking about! For it turns out that "Fizeau's relation"—which Einstein derived from special relativity and touted as a verification of his theory—can actually, once again, be obtained equally well from classical physics, a derivation which had in fact been carried out by the illustrious Henrick Lorentz himself.[19] As Bennett observes: "Unfortunately the denial of multiple causes for observed results is one of the key factors in current scientific rhetoric."

Leaving aside cases of this kind, what can one do to navigate a scientific literature which demands a high level of technical expertise in multiple fields? I propose to pursue a single trajectory based upon three interconnected experiments, which proves, I believe, to be definitive. It begins with the 1913 Sagnac experiment, in which an interferometer, mounted on a rotating platform, splits a beam of light, so that, with respect to the platform, one beam rotates clockwise and the other counterclockwise. As Bennett explains:

> The time for the counter-rotating light to circle the ring is less than when stationary, so this beam is superluminal. The co-rotating beam takes a longer time to traverse the

18. *The Meaning of Relativity*, 27.
19. *Galileo Was Wrong*, 446.

circle, so its speed is subluminal. In either case the speed of light exhibits anisotropy contrary to Special Relativity.[20]

The likely objection is that inasmuch as we are dealing with a "rotating" reference frame, special relativity does not apply. It happens, however, that Ruyong Wang *et al.* conducted an experiment in 2003, in which "the Sagnac effect is also obtained on a two-way linear path, by reversing a light beam sent out on a straight line on a moving platform and measuring the difference in return time."[21] What the Wang experiment indicates is that the speed of light is *not* in fact c in every inertial reference frame, as Einsteinian physics demands.

This leads to the question: are there other experiments which confirm this so-called Sagnac effect, or do the Sagnac-Wang experiments stand alone? Now it happens that the highly sophisticated Global Positioning System or GPS functions as a laboratory ideally suited to test the findings of Sagnac and Wang. For inasmuch as GPS operates by way of microwave beams connecting a terrestrial source to a satellite some 24,000 kilometers above the Earth, it is to be expected that "relativistic effects"—if such do exist—will come into play. What then does one find? "In 1984," Robert Bennett informs us, "GPS technician D. W. Allan and a team of international scientists measured the same effect on light as Sagnac did"! To be precise: "Alan and his colleagues found that microwave beams sent to an approaching GPS satellite take 50 nanoseconds less time to reach the satellite than beams sent to a receding satellite.... Once again, we have confirmation that the speed of light is not the same for

20. Ibid., 396.
21. Ibid., 403.

all observers."[22] It happens, moreover, that this correction proves to be crucial to the correct functioning of the GPS: "Each GPS unit must, without exception, take into account the Sagnac effect," and the aforesaid 50-nanosecond difference is in fact "automatically built into the computer program for GPS."

However—unbelievable as it may sound—at this point a tale of deception begins to unfold: for not only is this contra-Einstein finding not acknowledged, it is artfully concealed. As Ruyong Wang and his associate Ronald Hatch testify: the Jet Propulsion Laboratory bases its calculations officially on a so-called "solar system barycentric frame"—*modified*, however, *so as to yield precisely the same results as an Earth-centered or so-called ECI reference frame!*[23] Now, in plain English, this is deceit. We have, at this point, drifted far from the textbook definition of science as "the unbiased quest of empirical truth"!

✦ ✦ ✦

IT REMAINS TO SAY AT LEAST A FEW WORDS REGARDING A question obviously relevant to "the empirical status" of Einstein's theories: how, namely, does Einsteinian physics stand in relation to quantum theory? Let me note in the first place that quantum physics is in a way the very opposite of relativity theory. For so far from being "ideology driven," it imposed itself upon a generally reluctant scientific community. Not therefore on ideological grounds, but evidently by force of necessity. Scientists accepted quantum physics for the simple reason that it proved to be the only theory capa-

22. Ibid., 565.
23. Ibid., 569.

ble of dealing with the newly-discovered "quantum phenomena." And from its inception the new physics func-
tioned as the magic key which enabled physicists to unlock
the secrets of the microworld, from the "algebra" of atomic
spectra to the behavior of fundamental particles. And since
quantum mechanics à la Heisenberg and Schrödinger
merges into classical mechanics as we ascend into the
macro-world—that is to say, let Planck's constant h tend to
zero—the question presents itself whether there exists a
modified or "corrected" quantum theory which similarly
merges into *relativistic* mechanics.

Now let me note at the outset that I have never concerned
myself seriously with that kind of physics: have not so much
as read a paper dealing with any part of it. I speak thus as an
outsider. There are two things, however, I can say with absolute certainty: first, that the effort to "marry" quantum theory with Einsteinian physics—which has been ongoing for a
very long time, has enlisted a galaxy of brilliant physicists,
and engendered some of the most dazzling examples of
mathematical wizardry the world has ever seen—has failed
abysmally to achieve its objective. And I vividly recall
Michio Kaku, in a documentary film,[24] stretching his arms
wide and raising his voice as he informs us that relativity
and quantum theory differ ultimately *"by one hundred
twenty orders of magnitude!"* I don't understand what exactly
this means; but judging by the expression on Michio Kaku's
face, it must be about as bad as such things can get.

But no one should actually be surprised. I ask myself:
what has Einsteinian physics ever accomplished that one
should search for a *relativistic* quantum theory in the first

---

24. *The Principle.*

place? And in fact I regard it as yet another triumph of quantum physics that it has spurned the proffered union.

✦ ✦ ✦

IT SHOULD HARDLY COME AS A SURPRISE THAT CONTEMporary astrophysics—based as it is upon the general theory of relativity—has not fared too well. The grand expectations, fueled by the mystique of a "four-dimensional spacetime" in which moving rods contract, clocks slow down, and the continuum itself curves in unimaginable ways, have not materialized: the facts of observation seem not to be onboard. From the outset unanticipated difficulties have cropped up which could be "explained away" only by invoking the virtually unlimited powers of the most abstruse mathematics. When for example, in the first second of the postulated "Big Bang," the expansion proves too slow, the resources of mathematical wizardry can provide us with something termed "inflation" to set the process back on course. Or when it turns out that there is not enough matter in the universe to produce gravitational fields strong enough for the formation of stars and galaxies, there are mathematical geniuses with a flair for particle physics who can make up the difference with something termed "dark matter." The requisite ingredient can in fact be supplied in numerous varieties and flavors: by the magic stroke of a pen there appear before us higgsinos, photinos, gluinos, quark nuggets, and many many more such marvels to meet our needs. Never mind that not one of these wonders has ever been detected: the fact that they are supportive of "Big Bang" theory is proof enough.

We need not multiply examples of this kind. What confronts us here is a strategy for keeping a scientific theory alive by means of *ad hoc* postulates, assumptions "picked out of thin air" for that very purpose. The technique had long served as a mainstay of Darwinist biology before it came to be employed in the service of Einsteinian cosmology. I find it an interesting question, pertaining to the philosophy of science, whether the end-product of such a process, assuming the sequence does converge, has yet even a scintilla of scientific validity: have we then, in other words, discovered a truth—or simply constructed a fantasy? We need not however resolve that issue in the case of Einsteinian cosmology: for by 1996, difficulties arising from empirical data came into view which no "*ad hoc* magic" could dispel.

The problem resides in an electromagnetic radiation field known as the CMB: let us pick up the story at this juncture. As Robert Bennett explains:

> The Cosmic Microwave Background is considered the most conclusive piece of evidence for the Big Bang by current cosmology. It is the isotropic radiation bath that permeates the entirety of the universe. Accidentally discovered in 1964, it was soon determined that the radiation was diffuse, emanated uniformly from all directions in the sky, and had a temperature of about 2.73 degrees Kelvin. It is now explained as a relic of the evolution of the universe.[25]

The CMB is in fact regarded by theorists as a picture of the universe some 300,000 years after the Big Bang. Its isotropy, moreover, is tempered by "small" random fluctuations which in fact are needed to explain the formation of stars: it is for this reason that cheers went up, literally, when it was first announced that data from the COBE satellite had con-

---

25. *Galileo Was Wrong*, 498.

firmed the existence of such statistical variations in the CMB.

What the cheering scientists did not know is that additional information regarding the CMB was in the offing, which would prove to be far less felicitous. For it happened that, not long thereafter, hints of an "axis" began to appear, spreading alarm within the astrophysics community: its initial designation as "the axis of evil" testifies to the fact that the phenomenon—if it were to prove real—is by no means propitious to the Big Bang. In hopes of putting these fears to rest, another satellite, named "Planck," outfitted with the most marvelous instruments—and lavishly protected against all conceivable "extraneous" radiation which might cause a non-existent axis to emerge—was put into orbit in 2009, at which point the scientists could do nothing more than wait and hope for the best. It was a "make-or-break" scenario; for if that "axis" were to prove real, no *ad hoc* hypothesis whatsoever could save the theory: it would in truth be the end of Big Bang cosmology. And the verdict did come in, and the dreaded axis was there, plain as day—as if drawn by the Finger of God.

Let us pause to understand why this discovery does in fact prove fatal to the Einsteinian cosmology. The fact is that Big Bang cosmology hinges upon the so-called Copernican Principle,[26] which stipulates that the cosmos is perfectly homogeneous when viewed on a sufficiently large scale. Besides exemplifying the Einsteinian denial of "design" to perfection, this Principle also proves indispensable to relativistic cosmology on technical grounds: for unless one postulates global symmetries which epitomize the very "design" Einsteinians are pledged to deny, it constitutes the one and

26. Also termed the Cosmological Principle.

only condition under which the pertinent field equations can actually be solved for the cosmos at large. The Einsteinians, therefore, had one chance, and one chance only, to arrive at a global cosmology; and when that axis, that fateful Line, appeared in the CMB, that prospect collapsed.

Yet there is more to the story: for it happens that the plane defined by that circular axis *coincides with the ecliptic of our solar system*; and for Einsteinians this constitutes indeed a worst-case scenario: for what was supposed to be an accidental "speck" within a galaxy—which itself is supposedly but an accidental speck in a universe bereft of order, bereft of design—this "accidental speck within an accidental speck" turns out to *define the global structure of the universe!* At the risk of sounding anthropomorphic, I surmise "the Old One" may have *smiled* when he traced that axis. Or perhaps—to venture yet another anthropomorphic surmise— may simply *have had enough!*

UNMASKING "ANTHROPIC COINCIDENCE"

BY THE MIDDLE OF THE TWENTIETH CENTURY, ATOMIC physics and molecular biology had arrived at a point of vantage from which the existence of man upon "planet Earth" appeared so vastly improbable as to be, in effect, miraculous; and to folks who despise miracles, this poses a problem.

The central mystery resides in the fact that an exceedingly fine balance of the four fundamental forces known to physics is needed to render organic molecules stable enough to exist, which means that the basic constants of physics need be almost "infinitesimally" close to their given values— prompting the question: *why?* "What is a man," asks Carl Becker, "that the electron is mindful of him?"

What confronts us here in this "mindfulness" is evidently one of the most compelling and irrefutable instances of *design*. To the unprogrammed mind, moreover, this fact is immediately apparent: one recognizes the imprint of *design* through a quasi-direct realization that the phenomenon in question *could not*—by any stretch of the imagination—be the effect of *horizontal* or so-called "natural" causes. One doesn't require a university diploma or so much even as a high school education: the natural light of human intelligence suffices amply to arrive at that conclusion in a trice.

When it comes to the contemporary scientist, however, that normal power of comprehension seems no longer to be operative: for him the question "why the electron is mindful of man" constitutes a problem that calls ultimately for some kind of "hyper-physics" yet to be conceived. So he sets about to construct scenarios of the most imaginative kind to explain—or actually, explain away—this problematic "mindfulness," oblivious of the fact that he is leaving out of consideration what is actually the crucial point: namely, that *it takes*, not one, but *two modes of causality* to make a universe. These hyper-physicists are searching therefore— with indomitable determination no less—for something which in principle *cannot exist*. No wonder the search continues interminably to expand while it leads nowhere: one might as well scour the mathematical woods in search of a rational square root of 2! And no wonder, also, that the quest has fragmented into various technical domains defined by a *modus operandi* of their own, each fantastic in its own way. Meanwhile, as library shelves continue to bulge, the mystery of the electron's "mindfulness" remains as incomprehensible to the scientific *periti* as it had been when the search began more than half a century ago.

By the time John D. Barrow and Frank J. Tipler published

their monumental survey,[27] this branch of "science" had burgeoned into a field almost rivalling the classical domains of physics. Having elsewhere presented an overview and analysis of this strange twentieth-century meta-physics,[28] I will restrict myself to what could perhaps be termed the mainline argument, leading step by wondrous step to that astonishing conception known as the "multiverse."

The first thing that presents itself to the contemporary mind when it comes to explaining origins is of course the idea of "evolution": the problem, in this instance, is to conceive how the physical universe as such—which supposedly forms the basis of everything—could itself "evolve." How could such a thing as the fine-structure constant, for example—a numerical constant of physics which happens to be 7.2973531×10^{-3}, failing which "we" would not exist—how could that constant have conceivably "evolved"? A brand-new conception, obviously, is called for at the very outset of such an inquiry. And here the ingenious idea presents itself that inasmuch as, *for us*, the universe is evidently not an unobserved entity, *this very fact* imposes certain conditions of a *physical* kind. And though, at first glance, that notion may appear to be tautologous—like saying "the universe is perceptible because it is perceived"—that seeming tautology proves actually to have teeth as a so-called "self-selection" principle.[29] It leads in fact to the first "anthropic

27. *The Anthropic Cosmological Principle* (Oxford University Press, 1986).

28. *Ancient Wisdom and Modern Misconceptions*, ch. 10.

29. The idea was applied, for example, by Copernicus in the recognition that the phenomenon of retrograde motion in planets could be explained by the motion of the Earth-bound observer. In the context of the anthropic coincidence problematic, it was introduced by Brandon Carter in 1974.

principle" named WAP, the "W" standing for "weak." In the words of Barrow and Tipler, it affirms that "the observed values of all physical and cosmological quantities are not equally probable, but take on values restricted by the requirement that there exist sites where carbon-based life can evolve, and by the requirement that the Universe is old enough for it to have already done so."

We will pass over the fact that our hyper-physicists are apparently willing to base their conclusions upon the Darwinist hypothesis at a time when the discovery of the genome has already *de jure* invalidated that nineteenth-century conjecture. But let us continue: this WAP proved in any case to be less than sufficient, and was succeeded ere long by an SAP—"S" standing for "strong"—which asserts: "The universe must have those properties which allow life to develop within it at some stage of history." Yet that new so-called "principle"—which once again presupposes the Darwinist scenario and in truth explains nothing—that principle too does not yet suffice: for the question obviously remains from where and by what means one could obtain a universe satisfying that "strong" condition. And out of the maze of the purportedly "scientific" literature dealing with this issue, what might be termed a consensus appears to have emerged: following the lead of Stephen Hawking, a considerable portion of the scientific elite have opted for an infinite ensemble of possible universes, termed the "multiverse," as the ultimate solution to the riddle posed by the existence of our world.

There is a certain logic in this Ansatz; for in light of the SAP the question becomes: how then do we obtain a universe satisfying that stipulated condition? And there are, basically, only two options: by an act of creation of course, *or* by a "roll of the dice"—which evidently demands an adequate supply of "universes" from which an "SAP universe" can be drawn, as it were, "by chance."

Now, all this is of course perfectly insane, and brings home the lengths to which scientists of repute are willing to go in this ongoing "war against design." One cannot but ask oneself: whence comes this deep-rooted aversion—this profound animosity, one is tempted to say—*to the very conception of God*?

6

THE EMERGENCE OF
THE TRIPARTITE COSMOS

WITH THE DISCOVERY OF QUANTUM THEory, the "Reign of Quantity" has entered its terminal phase. The shift from classical to quantum physics—which evidently is not compatible with the pre-quantum worldview—heralds the demise of that Reign: when the fundamental science stands in contradiction to the prevailing Weltanschauung, the latter is bound to give way.

Let us look at the matter more closely. Beginning with Galileo and Descartes, the "enlightened" portion of mankind succumbed to the surprisingly crude notion of a "clockwork" universe. Reformulated in the course of the nineteenth century in more abstract terms through the discovery of electromagnetism, it is in the schematic of Einsteinian physics that this mechanistic notion attained its most sophisticated form. Meanwhile, however, it came to pass with the advent of quantum theory that the idea of clockwork causality in whatever mode has been invalidated, once and for all, as the presiding paradigm: not even the dazzling mathematics of David Bohm—with its quasi-magical pilot wave—can controvert this fact, as we have seen.[1]

1. See p. 27 above.

The pre-quantum concept of a deterministic causality retains of course a certain validity in the classical limit—realized mathematically by letting Planck's constant h tend to zero—but fails irremediably on the plane of quantum theory.

But whereas the demise of determinism has been widely acknowledged, what has so far remained almost totally unrecognized is the fact that quantum physics demands[2] *two* supra-physical conceptions: one *etiological* and the other *ontological*. The first—what I term *vertical* causation—comes into view, as we have seen, in the act of measurement, and proves to be more basic and more powerful than the horizontal modes of causality known to the physicist; and the second, amazingly enough, is simply the affirmation of the actual world "in which we live, and move, and have our being": how strange that despite decades of almost desperate efforts to circumvent "quantum paradox," it has apparently never occurred to the quantum-reality theorists that such a world might actually exist!

These then are the twin recognitions which not only resolve the quantum enigma, but *de jure* terminate the Reign of Quantity. It matters little whether our scientific pundits recognize this fact or continue to espouse disqualified premises: like it or not, the cycle of history initiated by Copernicus and Galileo is now drawing to its close. When the foundational physics—which proves to be the most "astronomically" accurate science the world has ever seen—declares its own insufficiency and points beyond itself, the

---

2. It does so, not on the basis of quantum mechanics itself, but on the grounds of metaphysics—which is after all the reason why, to this day, "no one understands quantum theory."

opinions of the reigning *periti* pale into insignificance: whether they elect to come aboard or not, history continues on its course.

✦ ✦ ✦

HOW THEN DOES THE RESOLUTION OF "THE QUANTUM ENIGMA" impact our worldview? In the first place it gives us back a "world" that *can be* viewed! One needs to recall that since the publication of Newton's *Principia* in 1687, the scientific pundits have imposed upon us a "world" that cannot be *viewed* at all. What has saved us from annihilation, meanwhile, is the fact that no one on Earth actually believes what the scientific *periti* advocate on that score: not even the top physicists themselves! To all except possibly the impaired, the sky is still blue and roses red, except in those rare moments when one engages in scientistic speculations of a Cartesian kind: *then only* do we deny what otherwise we staunchly believe. Though well-nigh universally unobserved, the fact is that our scientistic Weltanschauung—to the extent that we have made it our own—has plunged us into a state of collective schizophrenia, an affliction scarcely compatible with sanity. The first effect, then, of the aforesaid resolution, is that it cures us at one stroke of this malaise, and renders it possible, in particular, to be a physicist while remaining perfectly sane.

My second point relates to Richard Feynman's remarkable dictum: "*No one understands quantum theory.*" It happens namely that the resolution of the quantum enigma enables us not only to "understand quantum theory," but to understand at the same time why previously "no one" *could*. The point is that the so-called quantum world cannot be comprehended without reference to the sense-perceptible or *corporeal*: the very realm the existence of which physicists have long been taught to deny. The reason, moreover, why the quantum world cannot be understood "on its own" is quite simply that *it does not in fact exist*: for as Heisenberg was the first to observe, that putative "world" consists of so-called particles which are not in truth "things," but *potentiae*, which do not "yet" exist.

But not only has it now become possible to "do physics" while remaining perfectly sane—and even to "understand" what one is doing—but now, by virtue of these recognitions, one can actually *do so better*. Thus we have come to understand not only what physics *can* do, but also what it can't: for it happens that the ontological interpretation of quantum theory has uncovered a hitherto unknown mode of causation more fundamental than the causality indigenous to the physical world. For as the resolution of the measurement problem has brought to light, *vertical causation trumps horizontal*: has power, namely, to abrogate or "re-initialize" the Schrödinger equation. And this is a game-changing discovery, for it tells us that quantum physics is not in truth the absolutely "fundamental" science one has taken it to be, but is in fact restricted in its scope to an "underworld" of mere *potentiae*: that so far from being a "theory of everything," there is rather point in saying that it is, in a sense, a theory "of nothing at all."

I will mention in passing that students of Oriental philosophy will not be altogether surprised: it has been known for a very long time by way of the *yin-yang* that existence cannot be reduced to a single principle: that "it takes two to exist." To put it in Aristotelian terms: it requires *hyle* plus *morphe* (the Latin *materia* and *forma*) to make a world. And herein resides the ontological heresy of modern physics, which has in effect sought to build a cosmos out of *hyle*—out of "matter"—alone.

The objection may be raised that physics has actually to do with *quantities*: that it constitutes after all "the science of measurement." True enough! But here again the Scholastics hold the key: "*Numerus stat ex parte materiae*"—"Number stems from the side of *materia*"—they declare. And this means that even if the physicist could know literally everything pertaining to the quantitative aspects of the cosmos, something absolutely basic and indeed "essential" would still elude his grasp.

Strange as it may sound to modern ears, what physics as such has left out of account is precisely the *active* principle of cosmic existence: the Aristotelian *morphe* corresponding to the *yang* side of the Chinese icon. And this "white spot in the black field" has now revealed itself on the quantum level in the form of *vertical causality*: it is something physics as we know it can neither comprehend nor ignore—because in fact that causality manifests the cosmogenetic Act itself.

✦ ✦ ✦

ALL THIS WAS PREDICTABLE FROM THE START. THE LAPSE into a fantasy-world began with the philosophy of René Descartes, upon which the current interpretation of our physics is based: by banishing the *qualities* from the real or "external" world, the French metaphysician has in effect cast out the *yang*-side of cosmic reality. We have said that "number" or *quantity* derives from the side of *materia*: it needs also however to be pointed out that the *qualitative* content of the world—what the Cartesians have downgraded to the status of mere "sensations"—stands on the side of *morphe*, of *form*. What is left, therefore, after the Cartesian intervention, is a kind of half-world which, in truth, as we have said, does not exist. Yet it is this semi-world, precisely, which

Newtonian or "classical" physics has made its own, its "universe" over which that physics holds sway.

What commends this scientistic claim, at least superficially, is the fact that physics is primarily concerned, not with what things *are*—or whether they even "exist"—but simply with *how they move*: its fundamental laws are *laws of motion* after all. What causes a thing to move, moreover, are generally other things, which act upon each other by way of a causal chain. One may think of it as a kind of domino-effect propagating through space, which constitutes what we have termed "*horizontal*" causality. It is actually surprising that this simple notion of causality has enabled the prodigies of prediction and control physics has wrought, persuading some of the brightest minds that it covers the entire ground; what on the other hand is not surprising in the least is that in fact it does not. Yet that recognition will take time to disseminate: physics at large has, after all, proclaimed for over four hundred years that the world is made up of Cartesian *res extensae* interacting in complex ways via *horizontal* chains of causality, and even the discovery of electromagnetism—"ethereal" though it be—did nothing to dislodge that inbred notion: the Newtonian "clockwork" paradigm may have been refined but was by no means abandoned.

The picture began to shift with the discovery of quantum physics, which disclosed the startling fact that *in the quantum world there are no res extensae at all.* Physics had finally descended, so to speak, to its own level, which proved to harbor *potentiae* in place of "actual" *things.* To put it in Scholastic terms: physics had now arrived at the level of *materia signata quantitate* (matter under the determination of quantity), between *prima materia*—which properly speaking has no "existence" at all—and the lowest ontological stratum within the cosmic hierarchy, what I designate the *corporeal.* It is thus to this pseudo-world of *materia signata quantitate* beneath the corporeal that the objects of physics in truth belong, a fact which becomes apparent the moment physics has been divested of its metaphysical fantasies: has been "de-Cartesianized," one could say. The problem, however, is that whereas quantum physics itself has rejected the Cartesian presuppositions, the quantum physicists have not—which is precisely why the so-called "quantum reality problem" has proved for them to be *de facto* insoluble.

What presently concerns us is the fact that at the very moment when physics "descended," so to speak, to its own proper level, a *non-physical* mode of causation has come into view. We need to understand *why* it takes a "non-horizontal" causality to "manifest" the quantum world. Now it happens that the reason is not far to seek: for if the act of measurement does indeed entail a transition from the *physical* to the *corporeal* plane, then horizontal causality cannot take us there: for as previously noted, a transition between ontological planes can only be accomplished "instantaneously,"[3] and thus by way of *vertical* causation. The fact is that this hitherto unrecognized causality functions in this instance as the connecting link between the quantum world of *potentiae* and the *real* or *existent* world of corporeal entities. It actually makes perfect sense! Our connection with the quantum world hinges, after all, upon the act of measurement, which consequently requires a mode of causation *not* effected by causal chains. The moment, therefore, when physics attained to its rightful ontological status, VC was bound to enter the picture as the connecting link between the *quantum realm* and the *corporeal world.* From a meta-

3. See pp. 26–27 above.

physical point of vantage, moreover, it can be seen that vertical causality is needed to supply the *yang* component of corporeal existence, which the quantum realm as such does not possess—in a word, that vertical causation is "form bestowing," and therefore inseparable from the cosmogenetic Act. One could in truth say that it pertains to the authentic "Big Bang" which acts—not "in time," some fifteen billion years ago—but in the very *nunc stans* in which *"God makes the universe and all things,"* as Meister Eckhart declares.

Referring to post-Newtonian physics, Sir Arthur Eddington observed that "the concept of substance has disappeared." What has actually "disappeared," however, are the imagined *res extensae* which "classical" physics had spuriously presupposed. It is this ontological rectification, moreover, that has enabled physics to enter finally into its own proper domain, which proves to be categorically sub-corporeal—"sub-existential" in fact—for the simple reason that it is comprised exclusively of "quantitative" elements. What is missing in that so-called quantum world, as we have noted, is the *morphe* or *yang*-side of the coin: and that is precisely what vertical causality supplies or brings into play in the act of measurement, and in so doing, "actualizes" the quantum world "in part."

Getting back to knowing also what physics "can't do," I would like to emphasize how important this proves to be: think of the time and treasure wasted in the search for entities which prove to be *predictably* fictitious! Think of the "Big Bang" fiasco, not to speak of the Large Hadron Collider at CERN—which it took 100 nations to finance!—missioned to detect an array of "super-symmetry" particles conceived to exist in "space-time." My point is that a little *ontological* insight—sufficient, for example, to unmask the

Einsteinian postulates—could suffice to forestall scenarios of that kind.

The deeper issue, however, is this: having discovered the existence of vertical causality, and attained some initial understanding regarding the manner of its action, can that "more-than-physical" knowledge be put to scientific use? Can it lead to discoveries of a *scientific* kind, and possibly to as yet undreamed-of applications? I regard this to be more likely than not.

✦ ✦ ✦

THE QUESTION PRESENTS ITSELF, IN PARTICULAR, HOW the discovery of vertical causality impacts the biological sciences: for inasmuch as the distinction between animate and inanimate entities stems from their substantial form—from the fact, namely, that the substantial form of an animate creature is of a special kind, referred to traditionally as a *soul*—it follows that life and its phenomena constitute actually an effect of vertical causality. Our present biology, on the other hand, has eyes at best for the corporeal, which is to say that it views a living plant or animal as simply a "super-complicated" structure, which in the final count it conceives in *physical* terms. To the eyes of the contemporary biologist, it is thus ultimately the "tons" of specified information in the DNA that accounts for the observable phenomena: from the prevailing point of vantage there *is* after all nothing else that *could* enter the picture. Leaving aside the question how that "astronomical" complexity, embodied in that DNA, may have originated—which of course one is wont to answer in Darwinist terms—contemporary biology views a living organism thus as in effect a machine. It consequently perceives a plant or an animal as something which can in principle be understood "without residue" in terms of physics alone: that, when all is said and done, appears to be the underlying tenet upon which contemporary biology is based.[4]

The first point to be made is that this reduction of the animate to the inanimate—of the living to the merely "complex"—so far from being based upon scientific fact, is actually a groundless assumption, which gains strength from the fact that it is beyond our means to grasp whatever it may be that distinguishes the two. It amounts to saying that the living organism can only be what our methods of inquiry are in principle capable of detecting: that it must consequently reduce to a physical object, and that, at bottom, biology is no more than physics. But not only is this assertion unfounded, but as we have also come to recognize, it is in fact false: with the discovery of vertical causality the picture has radically changed. We now know not only that VC exists, but that it is in fact productive of corporeal being: not even a pebble can exist without a corresponding act of vertical causation. And this is a fact physics as such cannot grasp, let alone explain. The *a priori* notion, therefore, that physical science can account without residue for the phenomena of life—that living organisms reduce ultimately to a physical system—has thus been falsified: how can VC be left out of account in a living creature when it enters the picture even in a drop of water or a grain of sand?

The crucial question is whether the VC productive of a living creature can "override" the laws of physics, can in certain ways transgress these laws. To put it in Laplacian terms:

---

4. I do not wish to imply that such is the belief of *each and every* biologist; but though there may be exceptions, such appears to be indeed the rule.

if we knew the morphology of a living organism perfectly at a given moment of time, could we then *in principle* calculate its behavior with perfect accuracy? Or taking "quantum effects" into account: can the behavior of a living entity "violate the laws of physics," be they classical or quantum-theoretic? What we are asking in effect is whether a living organism is actually *more* than a physical system; and according to the currently prevailing view—the prevailing *assumption*, to be precise—it is not. Yet in truth there is not, nor in fact *can there be* scientific evidence of any kind in support of that assumption. The very idea of regarding a living organism as a "physical system" proves to be incongruous, inasmuch as it leaves out of consideration the very essence of a living organism, which resides after all in its substantial form and manifests through acts of vertical causality. This soul-generated VC constitutes in fact the life-force or *élan vital* of the organism, which both "produces" its body or "corporeal sheath" and renders it *animate*.

I say this, of course, not on scientific but on metaphysical grounds, inasmuch as the very question proves in fact to be metaphysical. But so too does the aforesaid reductionism of the contemporary biologist! The claim that the phenomena of life can be understood, without residue, in purely physical terms, so far from constituting a scientific fact, proves thus to be but another scientistic myth: a misconception which, under the banner of "science," has befuddled not only the scientific *periti*, but Western civilization at large. What has rendered the public at large vulnerable in that regard is the vast store of bona-fide scientific knowledge concerning the morphology and physiology of living organisms contemporary biology has in fact brought to light. One forgets, however, that these discoveries do not begin to validate the reduction of the animate to the inanimate: one can arrive at the same bona-fide scientific knowledge *without* assuming that living organisms reduce to a molecular machine. For not only is this reductive tenet unprovable on scientific grounds, it turns out to be scientifically useless as well: so far from opening empirical doors, it does no more than shrink our field of vision.

What in fact differentiates the animate from the inanimate is the vertical causation emanating from a living organism's substantial form, which could however have no effect at all if it did not, in some ways, impact its physical dynamics: for so long as every constituent particle of the organism functions strictly according to the laws of physics—be they quantum theoretic or classical—there can *be* no such effect. The VC productive of an inanimate corporeal entity, on the other hand, does not thus impact or "override" physical causality; for as we have seen, it constitutes the active principle which imposes these very laws upon the given entity. The laws of classical physics, one might say, are in fact determined by the substantial forms definitive of corporeal entities. One must remember: "laws" derive from the side of *morphe* as opposed to *materia*. And if it be indeed the case that the vertical causality productive of the corporeal domain "actualizes" the laws of physics—as our interpretation of quantum theory implies—it is readily conceivable that the substantial forms definitive of the biosphere will in certain ways modify or override these "physical" laws. The phenomenology of living organisms suggests as much: for it proves indeed to be a salient characteristic of living creatures to counteract the dissipative and "equalizing" tendencies operative in the inanimate world.

Having suggested that the substantial forms productive of the biosphere (what is normally termed a "soul") may be capable of "overriding" the laws of physics, it is time to recall that *in fact they do*. For as we have had occasion to see in the case of a human being, there are indeed scientific grounds in support of the tenet that man is endowed with what is traditionally termed "free will," which entails that he is capable of actions based upon *vertical* causation. The fundamental axiom of contemporary biology—the supposition, namely, that a living organism reduces to a physical entity—has thereby been disproved.

The fact is that our present-day biology is restricted in its scope to the physical substrate of a living organism: its "outermost shell" or cadaver one can say. Meanwhile these biological sciences continue to advance by leaps and bounds—without however coming one step closer to understanding what it is that differentiates a living creature from an inanimate entity: what, in other words, "enlivens" or animates a plant or an animal. At bottom we know neither what life is nor how it functions: we are acquainted only with its morphological and physiological feats. And this entails, let me add parenthetically, that the door to the paranormal and indeed the "miraculous" has not been closed.

✦ ✦ ✦

A FEW MORE WORDS REGARDING THE VEXED QUESTION OF "free will" may therefore be in order: for inasmuch as the behavior of living creatures is not fully determined by the laws of physics, it can be said that even an amoeba possesses that attribute in some degree. The tiniest living organism, thus, is more than an automaton, incomparably more than a mechanism driven by forces satisfying the equations of physics, as we have been taught to believe. Such forces "obeying the laws of physics" do of course enter the picture—but only in a peripheral capacity: they apply so to speak to "the outer shell," the cadaver as we have said. One catches a glimpse of this fact when one observes a living creature at the moment of death, when the soul separates from the body: whereas all the marvelous morphology is still in place, that body no longer functions as a living organism.

Getting back to "freedom of the will," it is to be noted that this capacity is something worlds removed from the "non-automatic" behavior of an animal: for that kind of "freedom" applies—not to amoebae or insects—but to man alone. And this brings to light another fundamental fact our biology fails to discern: there is a profound *ontological* distinction, namely, between the soul or *anima* of an animal and that of a human being. What renders us human is not simply a "soul," but what is termed a *rational* soul, which is something more, something incomparably greater. And this fact alone, let us note, not only falsifies Darwinism of any stripe, but brings to light its venom: for in thus depriving man of what is in truth "the image of God"—what Meister Eckhart refers to even as a *Vünkelin* or "little spark" of the Logos, of God Himself as it were—these modern-day biologists reduce him to the status of an animal.

The point to be grasped is that the biosphere is not to be conceived as an unbroken continuity, which is to say that there exist gaps between species and genera which no amount of horizontal causation can bridge. The fact is that the so-called "tree of life" exhibits an architecture, a hierarchic structure, with the *anthropos* at its very peak. Like the cosmos at large—with the Earth at its center—the biosphere exemplifies *design,* which is something a science based exclusively on horizontal causality simply cannot comprehend. I would note, therefore, that under these auspices Darwinism is *de facto* unavoidable, which explains why the scientific community refuses to acknowledge the fact that it proves to be scientifically untenable. As Ernest Mayr said in reference to calculations establishing the astronomical improbability of evolutionary origin in the case of an eye: "We are comforted by the fact that evolution has occurred."[5]

What impedes contemporary biology categorically and drastically restricts its purview stems from its failure to recognize the existence, necessity, and function of vertical causation; and inasmuch as the animate stands above the inorganic, that deficiency proves to be all the more debilitating. The great challenge, now, is to rectify these deficiencies; and whereas the prospect is doubtless challenging in the extreme, it can hardly be declared impossible.

5. Phillip Johnson, *Darwin on Trial* (Downers Grove, Il: Intervarsity Press, 1993), 38.

7

THE PRIMACY OF
VERTICAL CAUSALITY

VERTICAL CAUSALITY WAS IDENTIFIED IN THE context of quantum measurement: as the mode of causation, namely, which effects the transition from the physical to the corporeal domain. There exist other physical facts, moreover, that prove to be likewise indicative of VC: quantum entanglement, for instance, and the associated phenomenon of non-locality. When the measurement of an attribute of a particle at a point A *instantly* affects the state of its twin particle at B—conceivably light-years distant from A—the causality at issue *cannot* but be vertical. What gives rise to quantum entanglement, more-over, are interactions between the wave function and a *corporeal* entity, a fact which, once again, confirms that *vertical causality acts upon a quantum system by way of the corporeal plane*: that it "descends," so to speak, from the corporeal to the physical.

It behooves us now to "step back" and look at the etiolog-ical picture from a strictly ontological point of view. As stated at the outset, based upon what some have termed *cos-mologia perennis*, the integral cosmos proves to be *ontologi-cally tripartite*, and can be represented iconically by a circle, in which the circumference stands for the *corporeal* world, the center for the *spiritual*, and the interior for the *intermedi-*

ary.[1] It needs moreover to be clearly understood that these three components of the iconic circle correspond in truth to ontological domains of the integral cosmos which can be specified in terms of two *bounds*: the corporeal by space and time, the *subtle* or *intermediary*[2] by time alone, and the *spiritual* or *celestial* by the fact that it is subject to neither bound. What in the iconic representation appears thus to be the least of the domains—inasmuch as it has neither spatial extension nor temporal duration—proves thus to be actually the greatest: the ontological domain which in truth encompasses both the corporeal and the intermediary realms.

It is worthy of note, moreover, how drastically even our truncated cosmos, as conceived in contemporary cosmology, has actually *shrunk*; and one might add that the concomitant explosion in both its temporal and spatial extension—as measured supposedly in billions of years and light-years—merely exacerbates its indigence. Despite the official bluster, the fact remains that, compared to the integral cosmos contemplated perennially by the wise, that "brave new universe" amounts to little more than a speck of dust. Supposedly measuring billions of light-years, it is in truth too indigent to be envisioned at all: to do so one needs first to embellish that postulated universe with attributes it is actually unable to possess.

✦ ✦ ✦

A PHILOSOPHICALLY COGENT COSMOLOGY CANNOT BUT BE

1. It is vital to note that these three basic components of the cosmos have their counterparts in the *anthropos*, the human microcosm, which divides analogously into *corpus, anima*, and *spiritus*.

2. This is what has been termed the "astral plane" in occultist circles, and what in Orthodox Christianity is referred to as the "aerial world."

based on the recognition that the cosmos originated in a cosmogenetic Act which perforce is supratemporal, for the simple reason that time pertains to the cosmos alone: "*The world was created, not in time, but with time,*" St. Augustine observes. "Prior" to this inscrutable Act, nothing whatsoever existed—except of course the Universal Cause of all. Yet, surprisingly perhaps, this way of looking at cosmogenesis is yet somewhat incomplete: for according to the Aristotelian-Thomistic ontology—its so-called hylomorphism—something does in a way "pre-exist" the effect of the cosmogenetic Act: i.e., a certain "receptivity" or "recipient," if you will, termed *hyle* or *prima materia*, which is receptive of *morphe* or *form*, and in conjunction with *morphe* gives rise to the actual universe.[3] And that primary *morphe*, united thus to *materia*, gives rise to *substantial form*, which is what bestows reality upon a cosmic entity, what both causes it to *exist* and enables it to *act*: i.e., to deploy a *vertical* causality of its own.

This brings us finally to the fundamental questions: whence then does *horizontal* causation arise, and what constitutes its field of action? Now the answer to the first is virtually self-evident: *What gives rise to **horizontal** causality can be none other than **vertical** causation.* And as to its field of action, this too is not hard to discern: *The sphere of action of a horizontal cause can be none other than the corporeal domain*, for the simple reason that horizontal causation entails a transmission through space, and it happens that the corporeal world constitutes the one and only *spatio-temporal* domain; we must bear in mind that the spatial bound ceases to apply "above" the corporeal plane.

---

3. This corresponds evidently to what *Genesis* terms "the waters" in 1:2.

It needs however to be noted that horizontal causality does also in a sense act "below" the corporeal plane, that is to say, on the quantum level,[4] where it applies to entities "midway between being and non-being."[5] For whereas these entities may indeed be "less than real," the associated horizontal causality—as represented, say, by the Schrödinger equation—proves to be the most mathematically accurate physics mankind has ever known. What confronts us here is a "real" causality governing "less than real" entities.

Having shown that horizontal causation—whether acting on the corporeal or on the physical plane—constitutes an effect of vertical causes, it remains now to consider whether the latter are perforce cosmogenetic, or whether VC arising from a substantial form can be likewise productive of horizontal causality. To this end we need to take into account an obvious distinction: it is one thing to give rise to horizontal causality "*ex nihilo*," as it were—and quite another to *affect* or impact an existing causal chain (for example, by "re-initializing" the wave-equation). It thus becomes apparent that whereas it must be the cosmogenetic Act itself that actually gives rise to horizontal causation, secondary VC has nonetheless power to act upon existing chains of horizontal causality, to affect and alter them as we have had occasion to note: the vertical causality emanating from the substantial form of a pebble, for instance, has power to impede its multilocation. And let us not fail to recall at this juncture what we have said in the preceding chapter in reference to biology: that when it comes to substantial forms definitive of the animate order—the kind termed an "*anima*"—this capacity

---

4. As I have argued elsewhere, the corporeal and the physical domains share the same space and time. See *The Quantum Enigma*, 40.

5. In Heisenberg's famous words.

to "override" what we term the laws of physics is in fact definitive of the biosphere.

✦ ✦ ✦

GETTING BACK TO THE SO-CALLED QUANTUM WORLD, LET us remind ourselves that it constitutes a *materia signata quantitate*: a realm midway between *prima materia* or "pure receptivity" and corporeal being, endowed with purely *quantitative* attributes. What is it, then, that could have effected its formation: by what process of causality was quantity "added" as it were to *prima materia*? This is still something of a mystery. Here the *cosmologia perennis*, which has served as our guiding light up to this point, is of little help, inasmuch as nothing remotely resembling the quantum realm was ever conceived in pre-modern times. The crucial question is whether John Wheeler was right when he claimed that physics deals ultimately with a "participatory universe": that in some way this *materia signata quantitate* is actually "constructed" by the *modus operandi* of the physicist. There appear moreover to be fairly cogent grounds to conclude that such may indeed be the case: it happens that the pioneering work of Sir Arthur Eddington, which entails as much, has been strikingly confirmed in recent times by a physicist named Roy Frieden, who should by right be far better known than he is.[6] I would point out, moreover, that the notion of a "participatory universe"—the idea that the physical universe is in a sense "constructed"—is strongly supported by the recognition that corporeal entities can indeed "act" upon the subcorporeal realm by way of vertical

---

6. On this question I refer to *Ancient Wisdom and Modern Misconceptions*, ch. 2.

causation. And I will add parenthetically that an inquiry into that hypothetical "construction of the physical universe" *in light of vertical causality* strikes me as the ideal—and perhaps even as the only viable—starting point to arrive at a deepened understanding of physics: an understanding, namely, which brings to light the function of vertical causality, its intrinsic connection with horizontal causation. Yet be that as it may, on metaphysical grounds the fundamental fact remains that the quantum realm—like every other ontological domain—constitutes in the final count an effect of vertical causation.

What then is the *raison d'être* of *horizontal* causality: what constitutes its cosmic "function," so to speak? Now the salient characteristic of horizontal causation is that it has to do exclusively with *quantities*: with the *quantitative* side of cosmic existence. Not with quantity *per se* however, but with a certain kind of quantity: the kind, namely, which is inherently *spatial* and hence indigenous to the corporeal domain. And that, I surmise, is the kind referred to as "*numerus*" in the Scholastic dictum "*numerus stat ex parte materiae*" ("number derives from the side of *materia*"): the kind, namely, which measures or "metes out" spatial extension. And let us note that this is the very bound which defines the corporeal domain: one might refer to it as "extensive" quantity—the kind expressed by real numbers—as distinguished from quantity defined by integers and their ratios, which unlike the former has a significance by no means restricted to the corporeal domain.[7] Moreover, inasmuch as *time* (in

---

7. That is the reason why music, for example, with its harmonic and rhythmic ratios, can literally "lift us out of this world." The quantities in question here—significantly termed "rational"—are in a sense "qualitative," and pertain to a branch of mathematical science that may be

the sense of duration) is likewise measured by way of spatial extension, it too can be expressed in terms of real numbers.

Which brings us to the crucial point: *horizontal causality, as known to physics, assumes the form of a differential equation relating spatial and temporal magnitudes, and reduces thus to a "law of motion."* It needs however to be recognized that this law itself is imposed by an act of *vertical* causation: to claim otherwise is to put the cart before the horse. Given that the corporeal world originates in an ontological domain where there *is no* spatial extension at all—and where consequently the equations of physics do not apply—it follows that the spatial and temporal bounds, which these equations presuppose, cannot themselves be the result of horizontal causes. It can thus be affirmed as a theorem of authentic cosmology that *the fundamental laws of physics—expressive of horizontal causality—are based on vertical causation.*

We see from these cursory reflections that the scope and efficacy of horizontal causality within the integral cosmos turns out to be quite limited: not only is horizontal causality subsidiary to vertical, but its sphere of action is restricted to the corporeal and subcorporeal domains, what might symbolically be termed "the lower third" of the integral cosmos. To which one should add that as one ascends the *scala naturae* within the corporeal domain itself, the efficacy of horizontal causality is progressively diminished through the incursion of vertical modes. In terms of the traditional

characterized as "*harmonic*," which moreover played a decisive role in antiquity in support of the arts, the sciences, and even the spiritual life. All but forgotten in modern times, I surmise it will be rediscovered after the "reign of *numerus*" comes to an end. It richly deserves to be: for even as qualities trump mere quantity, so does the science of "harmonic" numbers outrank our contemporary mathematics in the larger scheme of things.

"mineral, plant, animal, and anthropic" partition, it appears that the hegemony of horizontal causation is restricted at best to the "mineral" or inorganic domain.

To explore the implications of this fundamental etiological fact for the sciences, for philosophy, and above all, for an understanding of man, of his origin and destiny—this is a task we leave for others to take up.

8

PONDERING
THE COSMIC ICON

From the point comes a line, then a circle.
SHABISTARI

INSCRIBED REPUTEDLY OVER THE PORTAL OF THE
Platonic Academy was an injunction which read: *"Let no
one ignorant of geometry enter here"*—what are we to
make of this? To begin with the obvious: it points to a connection between *geometry* and *metaphysics,* and suggests
that the former—as conceived in the Euclidean tradition—
may serve as a key to the profundities of the latter: a veritable
sine qua non in fact, as the words *"let no one"* seem to imply.

It behooves us, first of all, to recognize that for Plato all
"science," inclusive of metaphysics, was ultimately a matter
of *"seeing"*—but not simply with these, our corporeal eyes.
What is called for, evidently, is a seeing with the "eye of the
intellect," the authentic *nous* as distinguished from *psyche.*
There is however a connection between the two levels: the
"seeing" with corporeal eyes has a role to play also in the
metaphysical realm. And even in the case of ordinary
vision—as we have had occasion to note[1]—the actual "seeing" takes place "above time," and thus indeed on the plane
of intellect. Which brings us to the crucial point: just as

1. I am referring to James Gibson's discovery. See ch. 4.

visual forms can facilitate the perception of corporeal entities, so too can they catalyze an intellective perception pertaining to the metaphysical realm. An object of visual
perception, in other words, can serve as a sign pointing to a
metaphysical referent. A "metaphysical icon," then, is a visible form which—by virtue of an invisible correspondence—
can enable the intellective perception of a metaphysical
truth.

Getting back to the reputed inscription: one may thus
surmise that it refers to the use of certain geometric figures,
which we shall term "Euclidean," in an iconic capacity, as
the preferred means of entry into the metaphysical realm.
Judging by the inscription itself, this may indeed have been
the means *par excellence* in use at the Academy to "open the
eyes" of a novice to metaphysical vistas of which we are normally oblivious.

At the risk of digression, I would point out that there is at
least a partial analogy here with the use of mathematical
formulae in contemporary physics, which likewise constitute visual signs or symbols in support of an intellective act.
It might not be too much to suggest that the truly insightful
and creative practice of mathematical physics hinges upon
the mastery of this unrecognized art, which those called to
be mathematical physicists "pick up" somehow in the
course of their studies, and that the difference between a
Richard Feynman and the merely competent may have
much to do with how profoundly each can "read" the mathematical icons in question: how keenly he can *see* what the
equations indicate.

But back to Plato's Academy: Given that what might be
termed "Euclidean icons" play a pivotal role in the awaken-
ing of metaphysical perception, we need now to ask what it
is, precisely, that renders a geometric figure "Euclidean." But

what else could this be than the fact that it is constructed by means of the so-called Euclidean instruments: the straight-edge and the compass. The point is that this construction "enters the picture," as it were, and thereby bestows upon it a dynamic aspect in addition to its static form. A circle, thus, becomes something more than a closed circular arc: it comes to be perceived, rather, as a circular arc *swept out* by a compass. And even as we normally add a third dimension to a photograph or a drawing to "see" a depth implied but not given, so apparently it is possible, in the case of a Euclidean icon, to "see" intellectively the act of the compass of which the circle is but the result. Such an ability to see *more* than is visibly portrayed *must* after all come into play if we are to read our "circle cum center" as representing iconically, not only all of space and all of time, but even the supra-temporal realm itself, as given by a single point.

Having thus specified what is meant by a *"Euclidean"* as distinguished from a merely "geometric" figure, let us ask which, among all Euclidean constructs, is the simplest, say in the sense of involving the least number of steps to con-struct; and it is easy to see that this leads directly to what we term the cosmic icon. It is to be noted, first of all, that the initial step of every Euclidean construction is inevitably the determination of a "first point," called to this day "the ori-gin"; and this is something neither of the two Euclidean instruments can accomplish. The first step in every con-struction is thus to be effected by the geometer himself, without instruments, and thus as if by fiat. Inasmuch, moreover, as the application of the compass requires, not one, but two predetermined points, the second step is per-force to be carried out by the straightedge; and what the lat-ter determines or "metes out" is evidently a line segment OQ. The simplest Euclidean figure requiring both instruments in its construction is consequently a circle swept out by a compass centered at O.

It needs however to be understood that this centered circle constitutes but "the bare bones" of the cosmic icon: what renders the figure effectively iconic, as we have noted, is in principle the construction itself. It is the application of the compass, in particular, that brings into play the two cosmic bounds: of time by its single act—the sweeping out of the circle—and of space by its effect, which is to terminate the radii OP. It is thus by way of this seemingly innocuous instrument that one is enabled to accomplish the unlikely feat of "bringing time into the picture": all that is required, perhaps, to initiate this realization is a single clue, a single "hint"—which in fact the Master of the Academy himself provides: for instance in the *Timaeus*, when he refers to "time" as *the moving image of eternity.* Think of that compass with one point fixed at the center of the circle, while the other —its *"moving image"*—sweeps out the circumference; and behold: what presents itself to the qualified disciple is the *nunc stans*—the timeless "now that stands"— along with its "moving image."

But not only the *nunc stans* and the bound of time, but the bound of space presents itself as well: because at every "now that moves," P terminates the corresponding radius OP, and in so doing, imposes the spatial bound.

Of such a kind, then, is that "seeing," which no one on Earth however can bestow upon us. The Master, to be sure, can offer clues—"reasons" if you will—and perhaps even a little "nudge"; but when all is said and done, it is we ourselves who must draw that metaphysical vision forth from our own deepest center: the very center, namely, which itself stands "above time."

PONDERING THE COSMIC ICON

✦ ✦ ✦

FOLLOWING UPON THESE REFLECTIONS RELATING TO THE construction of the cosmic icon, let us take a look at the resultant figure itself to see what can be gleaned by the natural intelligence. What visibly confronts us are three elements or regions: the central point of the circle, the interior, and the bounding circle itself. And these three elements we have identified from the outset as representing, respectively, the spiritual realm transcending both time and space, the intermediary domain subject to time alone, and the corporeal subject to both the temporal and spatial bounds. Let us then, first of all, see for ourselves whether this reading accords with the construction.

Inasmuch as both bounds—temporal and spatial—are evidently imposed by the construction, and thus by the application of the compass, it is apparent, first of all, that the point O is subject to neither bound. As to the interior region, it is evidently swept out by the compass and is thus subject to the bound of time. And as to the third, represented by the circumference, it is thus a foregone conclusion that this represents the corporeal domain, inasmuch as there is now no other it could represent. Yet one also sees that "the moving point P" does double duty: on the one hand it visibly imposes the bound of time upon both the interior and the circumference, and on the other it differentiates the corporeal from the intermediary through the imposition of the spatial bound.

On closer examination, moreover, one discerns a kind of "iconic logic" built into the figure through the interplay of its static completion and dynamic construction as well as from the aforesaid "double duty" of P, which entails that a given element of the icon may carry more than one signification. Inasmuch as the icon implicitly identifies all such references, it literally entails *metaphysical* equations. And since every such equation determines a corresponding identification, the icon enunciates what may in truth be termed *metaphysical theorems*. I see four such implications.

THE AEVITERNITY
OF THE SPIRITUAL STATE

CONSIDER THE HIGHEST OF THE THREE REGIONS—WHAT we have referred to as the spiritual—and note that, viewed from the side of the construction, it is the fixed point O around which the rotating compass revolves: Dante's very *punto dello stelo a cui la prima rota va dintorno*. Its *prima facie* "eternity" is impacted thus by an apparent connection with time, in consequence of which it represents none other than *aeviternity*, precisely as St. Thomas Aquinas himself defines that conception when he writes: "*Aeviternity itself has neither before nor after, which can however be annexed to it.*"[2] Now this proves to be a crucial distinction. There are in effect two "kinds" of eternity which differ to the point of being incommensurable, but are commonly confused in our interpretation of the primary texts. A case in point is the "eternity" associated with what we have termed the "vertical" powers of the soul,[3] which is actually an instance of aeviternity. It needs to be realized that authentic eternity—as exemplified above all in the Kingdom of God—is something incommensurably greater than that.

2. *Summa Theologiae* I, Q. 10, Art. 5, corp.
3. See ch. 4.

THE PRIMACY
OF THE INTERMEDIARY

IT IS THE SWEEP OF THE COMPASS THAT TERMINATES THE intermediary domain, and in so doing, constructs the circumference representing the corporeal. The intermediary has therefore primacy with respect to the corporeal in that in principle it pre-exists the corporeal. One may thus conceive the corporeal domain as emanating from the intermediary through the imposition of the spatial bound.

THE PRIMACY OF TIME

NOTE THAT THE MOVING POINT P SWEPT OUT BY THE COMpass signifies two things: primarily it signifies an instant of time. But given that the circumference represents the corporeal domain, it likewise represents the corporeal domain in its entirety at that very instant. One might say therefore that *the moment of time imposes itself instantly upon all of space.* What we refer to as the "primacy of time" consists in this capacity to define an instantaneous "now" throughout the length and breadth and height of space—a feat, I would add, which cannot be accomplished by a transmission of any kind through space.

THE CYCLICITY OF TIME

THE BOUND OF TIME, AS DETERMINED IN THE EUCLIDEAN construction, is swept out by a rotation of the compass, a circular movement that returns to its starting point: and this fact cannot but be indicative of a corresponding property or characteristic of time, which we shall refer to as *cyclicity.* And this too, let us note, constitutes a theorem in the aforesaid sense, based upon the ambiguity of the initial point P, which proves to be likewise the endpoint of the circular sweep.

Let this much suffice to point out what I perceive to be the principal ontological theorems implicit in the cosmic icon.[4] And perhaps the first recognition we should take away pertains to the categorical difference between the bounds of time and of space: the fact, namely, that whereas time is permissive and dispositive of cosmic existence, space on the contrary is constrictive and terminating.

✦ ✦ ✦

HAVING ESTABLISHED THE AFORESAID THEOREMS, LET US reflect upon the significance of these metaphysical claims, beginning with the primacy of time. What is striking as one looks upon the cosmic icon is the fact that the interior, representing the intermediary region, visibly dwarfs the circumference representing the corporeal world. The very dimensionality of the respective domains ranks the intermediary region above the corporeal, which enters the picture as the one-dimensional boundary of the former. Add to this the fact that while there exists an ontological domain subject to time alone, there does *not* exist a region subject only to space. It thus appears that the spatial bound establishes the corporeal by terminating the intermediary domain.

I wish now to point out that this recognition enables us to literally *see* the invalidity of Einsteinian physics at a single glance: one need but look at a point P on the circumference of the cosmic icon and note that it represents both a

---

4. We are of course assuming the iconic "correctness" of what we have termed the "cosmic icon," something we neither prove nor justify.

moment of time *and* the corporeal world *at that moment of time*—an identity which disqualifies relativistic physics in its entirety at a single stroke. The cosmic icon enables us to *see* that there can be no such thing as "space-time": that when the cosmic clock "strikes" as it were, the resultant "now" is *instantly* defined throughout the length and breadth of cosmic space. The icon makes the case as sharply as it can conceivably be made: i.e., by representing a given instant of cosmic time and the corporeal cosmos in its entirety at that instant by one and the same point P. I find it ironic that a single point on the circumference of that icon suffices to disqualify the Einsteinian theory—and all that is built thereon—at a single glance.

As to the primacy of time, on the other hand, I would remind the reader that this has now been vindicated empirically in the phenomenon of non-locality, which in effect corroborates the existence of what we have termed the intermediary realm. Let us recall the scenario of the two photons X and X* in a so-called "twin state," which entails that a measurement of the polarization of one photon *instantly* determines the polarization of the other, no matter how great the distance separating the two particles may be. Here we have, first of all, an empirical exemplification of that ubiquitous "now" which the very notion of "space-time" negates. It is as if time does not "see" spatial separation, implying ontologically that what time does "see" can be none other than the intermediary domain itself. Or to put it another way: the "bond" connecting X and X*, inasmuch as it is not affected by spatial separation, pertains *ipso facto* to the intermediary domain; and this entails that both particles must in a way "pre-exist" on that plane.

There is thus a correspondence between corporeal entities and their "subtle" counterparts—a fact, let it be said, which leads in principle to the ancient Hermetic science known as *alchemy*. Suffice it to note in that regard that its two fundamental operations—known as *solve* and *coagula*—can now be scientifically conceived: for in view of the aforesaid correspondence one can pass, in principle, from a corporeal entity to its subtle counterpart pertaining to the intermediary domain by "lifting" its spatial bound (the alchemical *solve*), and can effect the reverse operation (the alchemical *coagula*) by the act of imposing such a bound.[5] What presently concerns us, however, is simply the fact that the phenomenon known to physics as "non-locality" entails the existence of the intermediary domain, and thereby corroborates in a way what we have termed "the primacy of time."

Having touched upon the subject of alchemy, I would add that it can be defined as the operative art or science (both terms are apt) of effecting vertical transitions within the tripartite cosmos, as the very name of its mythical founder suggests. For as one reads in the celebrated *Tabula Smaragdina* (commented upon by Albertus Magnus and translated by Sir Isaac Newton himself): "*And therefore I am called Hermes Trismegistus, because I own the three parts of wisdom pertaining to the entire world.*"

But getting back, for a moment, to the physics of Albert Einstein by way of closing that subject: we can summarize our findings in the observation that Einsteinian physics may

---

5. It is hardly necessary to point out that a science comes generally to be viewed as a so-called superstition once its operative principles are no longer comprehended; and in the case of alchemy, such has been the case in the West since the onset of the Enlightenment. Moreover, inasmuch as neither the *solve* nor the *coagula* can be effected by means of horizontal causality—and all knowledge regarding vertical causation seems to have vanished in our hemisphere—such a misapprehension was scarcely avoidable.

well prove to be the most *profoundly* erroneous theory ever seriously entertained, and doubtless the example *par excellence* of "mixing apples and oranges."

✦ ✦ ✦

FINALLY LET US RETURN TO WHAT MAY BE THE MOST PROfound teaching "rendered visible" by the cosmic icon: i.e., the *cyclicity of time*. What we know from the outset is that the key to the enigma resides in the paradigm of the compass: for the act by which it sweeps out the circle is itself cyclic in that the movement returns to its starting point.

Keeping our eye upon the "moving point P" in the construction of the cosmic icon, let us note, once again, that the "now" of time is inseparable from the "now" of the corporeal world at large. Clearly, one may interpret this to signify that *the corporeal world itself constitutes the "compass" that "sweeps out" time*, and that, by virtue of this fact, *time and the corporeal cosmos prove to be inseparable*. Gone is "the empty container" paradigm, that figment of a Cartesian imagination: it has now been replaced by "time" as an aspect or dimension of the corporeal world, which as such is not "empty" at all. What is it, then, that time "carries"? The best answer, perhaps, is that it is a certain power which rises and falls, conveyed by the cosmos at large as it "sweeps out" time.[6]

I find it striking how the cosmic panorama itself discloses this fact with the utmost clarity the instant we revert to a *geocentric* cosmography. Behold the cosmos, in its entirety,

---

6. Needless to say, premodern civilizations, virtually without exception, held a cyclic view of time. See, for instance, Robert Bolton, *The Order of the Ages* (Kettering, OH: Angelico Press/Sophia Perennis, 2015).

rotating diurnally—like a gigantic Compass—around the Earth's polar axis, meting out the cycle of days and nights! We are cognizant, of course, of the physical or "measurable" parameters caused by this diurnal rotation: my point is that there exist "non-measurable" (properly called "subtle") alternations as well, which likewise have their effect. And in addition to the cosmos at large rotating diurnally around the Earth, there are the planets—the "wanderers"—beginning with the Sun, meting out larger cycles with corresponding "seasons" of their own. The result is a kind of "music of the spheres" of which the contemporary scientist is totally oblivious—except of course for its electromagnetic and gravitational effects, which he can register the way a deaf man, say, can register music by means of an oscilloscope.

What post-Enlightenment man cannot grasp even remotely is that time owns not merely quantitative or "measurable" characteristics, but *qualitative* properties as well, which play a vital role in the economy of life. Wedded as he is to the physicist's conception of time, which is exclusively quantitative and based upon measurement, he is unable to realize that this identification of time with its measurable effects is no less fallacious than the Cartesian notion, say, of a *res extensa*. Whether our pundits of the Enlightenment realize it or not, "*there are more things in heaven and Earth, Horatio, than are dreamt of in your philosophy.*" There are, for instance, "seasons" carried by temporal cycles with a *qualitative* content of their own, which have their effect— whether we can measure it or not—especially upon living organisms, beginning with man himself. It makes perfect sense to speak of time as the carrier not only of quantitative but of *qualitative* determinants as well.

This brings us, finally, to the threshold of what is perhaps the deepest science ever revealed to mankind, the remnants of which have survived to this day under the caption "astrol-ogy," the reputed "science of the stars." No wonder the disci-pline has been reviled since the Enlightenment as a "pagan superstition"! For it *is* in fact "pagan" in the sense that it goes back to pre-Christian times, and literally a "superstition" as something "left over" which is no longer comprehended, no longer in truth understood.

Now, it is not my intention, at this juncture, to provide so much even as a glimpse of authentic astrology; this would require a treatise in its own right, which moreover I am by no means qualified to deliver. A few observations on the subject is all I am able to provide.

The first thing to be noted in regard to astrology is that the causality upon which it is based proves to be ineluctably *vertical*, and derives moreover from the highest reaches of the cosmos: the planets and constellations especially. And it brings man, the human microcosm, into the picture by way of his horoscope as defined by the position of the planets relative to the zodiac at the moment of birth. Now, the first misconception—generally upheld by friends and adversar-ies of astrology alike—is the notion that this horoscope refers to the *effect* of the aforesaid planetary configuration upon the soul: but such a one-sided picture is misleading, to say the least. One needs to understand, in the first place, that man does indeed constitute a "microcosm"—a universe in miniature as it were—which as such admits of an astro-logical description; and most importantly, that *the time and place of his birth is by no means arbitrary or accidental.* Whereas all are created in the "*omnia simul*"[7] of the Creative Act, we are born in time "when we ought to come into

7. Ecclus. 18:1.

being," as St. Augustine avers.[8] There are no "accidents" in that regard, no "rolling of the dice": the planets and constellations, at the moment of our birth, do actually no more than announce—in the precise language of astrology—*who it is* that has now made his entry into this world.

The next thing to grasp is that, contrary to what is generally believed, astrology is not, strictly speaking, predictive in respect to human affairs: what it tells us in reference to the future has to do with trends, what Shakespeare refers to as "tides in the affairs of men": there is no question here of an inexorable fate. On the other hand, by far the most accurate and potentially useful information astrology has to convey refers to who and what we are; and what above all matters is the stunning revelation that each of us is endowed with a unique constitution, expressible in astrological terms, which points to a likewise singular destiny as something given us to achieve.

In conclusion I wish to say a few words on whether the authentic claims of astrology accord with the teachings of Christianity, a question which seems generally to elicit more confusion than clarity. To begin with, I would point out that Scripture itself points to the existence of an authentic astrology, beginning with Genesis 1:4, which reads: "*And God said, Let there be lights in the firmament of the heaven to divide the day from the night; and let them be for signs and for seasons, and for days and years.*" It is clear that the "*lights in the firmament*" refer not only to the two principal lights, namely the Sun and the Moon, but to celestial "lights" in general, inclusive of planets and stars, all of which, moreover, are given us both for "*signs*" and for "*seasons*." To begin with the latter: this refers to what I have touched upon earlier by way

8. *De Genesi ad Litteram* I.2.6.

of explaining the "qualitative" content of time, which Gene-sis refers to here as "seasons." By the term *"signs,"* on the other hand, the verse alludes evidently to an astrology: for where there are "signs," there must in principle be an art or science to "read" these signs, to decipher them. And as a matter of fact, such a science is referred to in Scripture, here and there, as a historical reality: recall, for instance, the words of the *"wise men"* to Herod the king: *"Where is he that is born King of the Jews? For we have seen his star in the east, and are come to worship him."*[9] Most assuredly, Christ is no ordinary man! The point, however, is that if there were no such thing as a bona-fide astrology, the passage from St. Matthew would be void of sense.[10]

✦ ✦ ✦

THE QUESTION NOW OBTRUDES WHETHER THE CONCEPTION of a trichotomous cosmos—and perhaps even of the cosmic icon itself—is in some way sanctioned by the sacred litera-ture of mankind, beginning with the Judeo-Christian. While visibly pertinent texts appear to be both rare and hard to identify, it should be noted that such texts do exist, and prove as a rule to be enlightening. The first such refer-ence that comes to mind is the majestic passage in Proverbs:

9. Matt. 2:2.

10. I would mention that Malachi Martin—doubtless an authority in all things "angelic"—spoke of a "destiny angel" assigned to us at birth (although I have yet to meet a theologian who has so much as heard of this). My point is that there could hardly be a "destiny angel" assigned to us at birth if there were no such thing as a "destiny" associated with that moment. Let Christians who do *not* dismiss astrology as a "pagan super-stition" be reassured: in the final count they stand on incontestable ground.

"*He set his compass upon the face of the deep,*"[11] attributed to Sophia. In fact, it is she herself who declares: "*I was there when he set his compass upon the face of the deep.*"

It may be of interest to note that a somewhat similar text is found in the Rig Veda—possibly the oldest scripture in the world—which reads: "*With his ray he has measured heaven and earth.*"[12] There is a temptation moreover to connect this text to the cosmic icon by associating that "ray" with the first Euclidean instrument—but there appears to be little factual support for this conjecture. Certainly both the Hebrew and the Vedic texts refer to cosmogenesis, which they conceive as an act of measurement; but it is questionable that the Vedic verse contains something essential not given in the Hebrew.[13] What, on the other hand, proves to be directly relevant to the cosmic icon and enlightening in the extreme is the Gospel parable concerning the leaven "*hid in three measures of meal.*"[14] Here is the full text: "*The kingdom of heaven is like unto leaven, which a woman took, and hid in three measures of meal, till the whole was leavened.*"

It is apparent, first of all, that the "*three measures of meal*" correspond indeed to the three cosmic realms: the spiritual or aeviternal, the intermediary and the corporeal. The "*leaven,*" which Christ himself identifies with "*the kingdom of heaven,*" proves thus to be something inherently transcendent and ultimately divine, which gives reality—gives *being*—to the cosmos in its tripartite manifestation. Who

11. Prov. 8:27.

12. Rig Veda, viii, 25, 18.

13. Ananda Coomaraswamy has demonstrated that the notion of "measurement" is basic to the Vedic understanding of "creation." See his *Notes on the Katha Upanishad*, 2nd Part.

14. Matt. 13:33.

then, let us ask, is *"the woman"*? And clearly, she can be none other, once again, than Wisdom: the very Sophia who *"was brought forth"* when there were as yet *"no depths,"*[15] let alone the things of creation. It is she who *"hides the leaven"* in *"the three measures of meal."*

It behooves us now to reflect upon this "leaven": for it happens that much—everything in fact that ultimately counts!—rides upon that issue. We need in particular to ponder the words of Christ which interpret the "leaven" to signify the Kingdom of Heaven. I would note, in the first place, that as "leaven," this Kingdom is indeed *"within you,"* as Christ himself declares;[16] it is *"within"* you in the fullest conceivable sense, for as the parable tells, it is within each "sheath" of our tripartite being: the corporeal, the psychic, and the spiritual. Yet as the "leaven" added to *"three measures of meal,"* it is something other than *"meal,"* something thus distinctly *"not of this world,"* as Christ likewise affirms.[17] The decisive point is that the Kingdom of Heaven, though *immanent*, is yet radically *transcendent*. It thus supersedes not only the corporeal and the intermediary realms, but even the supreme cosmic domain represented by the central point of the cosmic icon: Dante's *"punto dello stelo a cui la prima rota va dintorno."* When a Christian speaks of "life *eternal*," he is thus referring to something *infinitely greater* than "life *aeviternal*," which yet pertains to the trichotomous world.

In declaring the Kingdom of Heaven to be immanent, the parable issues an open invitation, as it were, to strive after that Life, no greater than which can be conceived. We are,

15. Prov. 8:24.
16. Luke 17:21.
17. John 18:36.

each and every one, invited—by the Son of God Himself—to the enjoyment of that eternal Life, which He likens to a Wedding Feast.[18] And what should *de jure* melt every human heart is that in offering us this Life, the Son of God is actually offering Himself: for as He likewise declares, "*I am the resurrection and the life.*"[19]

The "*leaven,*" in the final count, proves thus to be *Christ Himself*: it is He that resides at the core of all being as the font of all that is good. Like the actual leaven in the meal, it is He that renders the cosmos hospitable and "flavorsome." In all that is good, He is the Goodness; and in all that is beautiful or sublime we catch a glimpse of His presence. As Meister Eckhart observes:

> If God were not in all things, Nature would stop dead, not working and not wanting; for whether thou like it or not, whether thou know it or not, Nature fundamentally is seeking, though obscurely, and tending towards God.[20]

In fine finali, the cosmos constitutes thus a theophany. And this is the Fact that trumps all other facts: the Secret the wise have had their eye upon ever since the world began. There are those who would go so far as to maintain that the cosmos actually has being only inasmuch as it manifests or mirrors God; as St. Augustine—gazing upon the things of this world—exclaimed to God: "*An existence they have, because they are from Thee; yet no existence, because they are not what Thou art.*"[21] Let us not however concern ourselves

---

18. Matt. 22:1–14.
19. John 11:25.
20. *Meister Eckhart*, trans. C. de B. Evans (London: Watkins, 1925), vol. I, 115.
21. *Confessions*, vii, 11.

with what the universe might or might not be "in itself": what counts is that it is *"leavened"* by Christ.

Now, what associates this literally *marvelous* parable concerning the *"three measures of meal"* with the cosmic icon is, first of all, the fact that, macrocosmically interpreted, it speaks of the universe as trichotomous.[22] The parable constitutes thus one of the rare *dicta* in Scripture which speaks—parabolically to be sure—of the cosmic trichotomy.

But there is a second fact which associates the parable unmistakably with our icon: it speaks namely of "three *measures* of meal." Think of it: by this designation it affirms that the three domains of cosmic existence were brought into being not simply by divine fiat, but specifically by an *act of measurement*. The reference to the cosmic icon could therefore be hardly more direct: for as we have noted, the three domains of that icon are likewise defined by a single act of measurement, which is moreover accomplished by means of a compass—in keeping with the words of Sophia: *"He set his compass upon the face of the deep."*

✦ ✦ ✦

ONE OF THE MOST EXPLICIT AND PROFOUND REFERENCES to the cosmic trichotomy in the sapiential literature is to be found in the Māndukya Upanishad, which is devoted exclusively to that issue. The Upanishad conceives of the three *loci* or "worlds"—the so-called *tribhuvāna*—as answering to three distinct modes of knowing, which correspond (in

22. In light of Patristic commentary, moreover, one can likewise associate that trichotomy with the microcosmic triad *corpus, anima, spiritus,* an association made for instance by St. Jerome. See St. Thomas Aquinas, *Catena Aurea* (Saint Austin Press, 1997), vol. I, 505.

ascending order) to the waking state, the dream state, and the state of dreamless sleep: what we access in these states pertains thus in principle to the corporeal, the intermediary and the spiritual worlds, respectively. I say "in principle," because such is not actually the case—notably because, in the state of dreamless sleep, we seem to access nothing at all! Let me interject, from a Christian point of vantage, that such is the effect of Original Sin: it is the Fall of Adam that has deprived us of access to the spiritual plane. As St. Paul explains, we have been truncated, as it were, at the level of *psyche*—deprived, in other words, of ingress into the spiritual world—and thus reduced to the status of a *psychikos anthropos*, someone *"who receiveth not the things of the Spirit of God."*[23] And so it has come about that when we enter into the state of deep sleep—pass through the gateway, as it were, leading into the spiritual world—we experience nothing at all. Such, however, is not the case when it comes to saints of high order, men and women who have advanced sufficiently in quest of God to regain access to the spiritual realm. As we read explicitly in the Bhagavad Gita: *"In that which is night to all beings, the in-gathered man is awake; and where all beings are awake, that is night for the renunciate who sees."*[24]

23. 1 Cor. 2:14.

24. Bhagavad Gita II.69. It should be noted, first of all, that the "in-gathered man"—the Sanskrit term is *"muni,"* which literally designates someone who (like the disciples of Pythagoras) observes a vow of silence—refers here to a renunciate who has risen to a state of spiritual vision. So too the words "is night" do not signify that the man of spiritual vision is simply ignorant of the two lower domains—which would be absurd—but means rather that he perceives them with a spiritual sight, and thus, comparatively, "as night." What "all beings" behold as the real is indeed "night" for him who "sees."

It will be instructive, now, to reflect upon the dream state, in which we enter into the intermediary domain as exemplified microcosmically by the *psyche*, which, as we have noted,[25] stands midway between the corporeal and the spiritual worlds in that it transcends the bound of space but not of time. Now this claim may strike us as incongruous, given that in dreams we evidently experience entities more or less like the things we perceive in the waking state, inclusive of their spatial boundaries. Yet even so, the objects perceived in the dream state are not in truth subject to spatial bounds—which is of course precisely what differentiates them from *corporeal* objects. They do not, in other words, exist *in space*: though perceived as spatial entities, they are in truth *subtle* (Sk. *sūkshma*) in that they pertain to the intermediary domain. Whosoever doubts this—whoever, in other words, imagines these objects to be corporeal—should try to ascertain the spatial coordinates of a mountaintop, say, perceived in a dream; he will discover soon enough that this proves to be categorically undoable. It is otherwise when it comes to the bound of time: the very fact that the ring of an alarm clock, say, may coincide with a given event or "moment" in a dream shows that the temporal constraint is still in effect.[26]

Having considered the correspondence between the "three worlds"—the *tribhuvāna*—and the associated states of consciousness in man, the Upanishad shifts abruptly from the "three worlds" to what it refers to simply as *turiya*,

25. See ch. 4.
26. Clearly, to understand what differentiates the "subtle" (Sk. *sūkshma*) from the "gross" (Sk. *sthūla*) it is needful to divest oneself of various customary assumptions: here too it is a question of "back to the facts"!

meaning "the fourth." We are told nothing further concerning this *turiya*, which is not even named but merely "pointed to" as literally the fourth in the given enumeration. We are told, in effect, that just as the subtle world transcends the corporeal, and the spiritual the subtle, so *turiya* transcends even the spiritual.

There is a certain correspondence, thus, between *turiya* and the "leaven" of the Gospel parable in that both transcend the ascending sequence consisting of the corporeal, the intermediary and the spiritual worlds. On closer examination, however, one sees that *turiya* not only transcends the *tribhuvāna*, but is immanent in each as well: like the "*leaven*," it is "hid" in the three "*measures of meal.*" Here is why: it follows, first of all, from what we have termed "*the primacy of the intermediary*," that the latter is immanent in the corporeal. So too it follows from "*the aeviternity of the spiritual state*" that the latter is immanent in the intermediary. And thus, finally, *turiya* would not in truth be "the Fourth" if it were not, in turn, immanent in the spiritual or "celestial" state. The correspondence between *turiya* and the "*leaven*" is thus complete: like the latter, *turiya* turns out to be both immanent and transcendent in respect to the integral cosmos.

The kinship does not however appear to extend beyond the properly metaphysical to the authentically Christian sense of the parable. I see in *turiya* no Kingdom of Heaven and no Wedding Feast, no Holy Trinity and no Incarnation: no actual union, thus, between man and God—except for that of the "*dewdrop*" said to "*slip into the shining sea.*"[27]

---

27. The metaphor is taken from Sir Edwin Arnold's *The Light of Asia*, a remarkable poem presenting the life and teachings of Gautama Siddhartha, the founder of Buddhism.

POSTSCRIPT

THIS BRINGS TO A CLOSE OUR BRIEF REFLEC-
tions on the scientific significance and implications
of vertical causality. It remains to point out that
having thus initiated what might broadly be termed *a rediscovery of the vertical dimension*, we may have prepared the
ground for a shift in the Weltanschauung of Western civilization. I believe, moreover, that a scientific *metanoia*, based
on a rediscovery of vertical causation, is apt to inaugurate a
cultural *metanoia* as well, which may "open doors" bolted
shut centuries ago.[1] Above all, it may enable men and
women to discern, once again, the *raison d'être* of human
birth, which is and ever shall be *the attainment of life eternal
in union with God.* We have been collectively distracted and
deviated for centuries from catching so much as a glimpse
of that God-given imperative: has the time perhaps arrived
when that descending arc of history will come to an end?
Such appears indeed to be the case.

That arc began in a very real sense with Galileo's hypothesized displacement of the Earth. It needs however to be
understood that this conjecture does not stand alone: for it
happens that the decentralization of the Earth goes hand in
hand with a corresponding decentralization of man. What
has in effect been lost are both the macro- and microcosmic
manifestation of that central point in the cosmic icon: that
"pivot around which the primordial wheel revolves." There are

---

1. On the relation between "science" and culture I refer to *Ancient
Wisdom and Modern Misconceptions*, especially the chapter entitled "Science and the Restoration of Culture."

in truth *two* centers: the macrocosmic and its counterpart in the microcosm, the *anthropos*; and the two centers are in fact inseparable. How, then, are they connected? And by now the answer cannot but stare us in the face: *that universal and transcendent Center of the cosmos is connected to its counterpart in man[2] by an act of vertical causality, which is none other than the cosmogenetic Act itself.* Neither spatial distance nor temporal duration, thus, separates *our* Center from that *"pivot"* around which *"the primordial wheel revolves."* And this, I surmise, constitutes the Mystery wise men have pondered ever since the world began: their Quest has ever been for that *"punto dello stelo"* hidden deep within the heart, which is *"the eye of the needle"* through which *"the rich man"* cannot pass, the *"narrow gate"* the *"pure in heart"* alone can enter.

How then did the Galilean intervention impact this Quest, this *longing*, however dimmed? It did so, ontologically, through the subjectivation of the qualities, and cosmographically, by the denial of geocentric cosmology. What remains, following these twin reductions, is on the one hand the phantasm of a clockwork universe driven by a horizontal causality, and on the other a de-centered humanity: for when the cosmos loses its center, so does the microcosm, so as a rule does man. The overall impact of the Galilean intervention proves thus to be twofold: on the one hand what René Guénon refers to as *"the reign of quantity"* engendered by Cartesian bifurcation, and on the other what might well be termed *"the reign of relativity"* symptomatic of a decentralized humanity in a decentralized universe. The congruity of God, man, and cosmos became thus newly compromised,

---

2. Which can be identified with his substantial form or soul.

and in consequence of this breach the *anthropos* himself has begun to disintegrate at an unprecedented rate: *the Galilean impact upon humanity could thus be viewed as a second Fall.*[3]

In light of these reflections it is evident that the impact of the Galilean revolution upon Christianity and Christian culture at large was in fact bound to be fatal. *Christian civilization has need of the pre-Galilean worldview*—and this fact was recognized from the start by those who had "eyes to see": think of the impassioned words of John Donne, penned in the year 1611, when the Galilean revolution had barely begun: *"And new philosophy calls all in doubt,"* he laments; *"'Tis all in pieces, all coherence gone,"* he cries! Yet no one has made the point more sharply than Herman Wouk when he proclaimed that Christianity has been *dying* "ever since Galileo cut its throat." I find it tragic that our contemporary theologians and churchmen seem, almost without exception, to have not so much as the faintest idea what Herman Wouk was talking about—which only goes to show, however, how profoundly right he was.

And this brings us finally to the crucial point: *in light of the facts delineated in this monograph, it appears that the Galilean arc of history is presently drawing to its close*: the rediscovery of vertical causation alone—along with the resultant unmasking of Einsteinian relativity—implies as much. For as we have come to see, the recognition of vertical causation opens the door to a rediscovery of the integral cosmos—the actual world in which we *"live, and move, and have our being"*—which not only exonerates geocentrism, but brings to light the existence and the ubiquity of the veritable Center.

3. On this issue I refer to "'Progress' in Retrospect" in *Cosmos and Transcendence.*

Let Christians—and all who bow before God—rejoice: the scourge of relativism and irreligion has now been dealt a mortal blow! Following four centuries of intellectual chaos and *de facto* incarceration within his own distraught psyche, *homo religiosus* is now at liberty, once again, to step out into the God-given world, which proves to be—not a mechanism, nor some spooky quantum realm—but its very opposite: a *theophany* ultimately, in which *"the invisible things of Him from the creation of the world are clearly seen, being understood by the things that are made, even His eternal power and Godhead."*[4]

4. Rom. 1:20.

INDEX

Albertus Magnus, St. 110
Aristotle 19–20, 32, 80
Augustine, St. 95, 114

Barrow, John D. 72–74
Bell, John Stewart 9–11, 28
Bennett, Robert J. 63–64, 69
Bohm, David 27, 77
Bohr, Niels 4, 7–8, 10, 12
Broglie, Louis de 27, 77

Cajal, Ramón y 41
Carter, Brandon 73
Copernicus 73, 78
Crick, Francis 39

Dante Alighieri 117
Darwin, Charles 48
Dembski, William 35, 37, 48
Democritus 2, 15, 25
Descartes, René 13, 15–16, 23–25, 45, 47, 77, 81
Donne, John 125

Eckhart, Meister 36, 118
Eddington, Arthur 16, 22, 84, 97
Einstein, Albert 7, 10–12, 48–68, 71, 110

Feynman, Richard 8, 79, 102
Frieden, Roy 98

Galileo, Galilei 45, 47, 77–78, 123

Gibson, James 15, 37–40
Gilson, Etienne 33
Gödel, Kurt 42–43
Guénon, René 124

Hatch, Ronald 66
Hawking, Stephen 74
Heisenberg, Elisabeth 5
Heisenberg, Werner 3–5, 9, 14, 16, 19, 38, 67, 80, 96
Herbert, Nick 11
Hermes Trismegistus 110
Herod 115
Hesiod 22

Kaku, Michio 67
Kelvin, see Thomson, William

Laplace, Pierre Simone de 27
Lewontin, Richard 57

Mach, Ernst 58–59
Martin, Malachi 115
Maxwell, J. Clerk 51–52, 54
Michelson, Albert 55, 57, 61
Morley, Edward 55, 57, 61

Neumann, John von 9, 11
Newton, Isaac 2, 10, 13, 17, 23, 45, 47, 52, 61, 79, 110

Paul, St. 126
Penrose, Roger 43
Planck, Max 1, 3

Plato 31, 40, 44
Popov, Luka 59

Sagnac, Georges 64–66
Schrödinger, Erwin 26–28,
   67, 80–81
Shabistari 101
Shakespeare, William 114
Stapp, Henry 11
Sungenis, Robert 63

Thomas Aquinas, St. 33, 106
Thomson, William 4, 19, 50
Tipler, Frank J. 72–74
Turing, Alan 41

Wang, Ruyong 65–66
Wheeler, John 97
Whitehead, Alfred North 12–
   14, 45
Wouk, Herman 125

ABOUT THE AUTHOR

WOLFGANG SMITH WAS BORN IN VIENNA IN 1930. At age eighteen he graduated from Cornell University with majors in physics, mathematics, and philosophy. At age twenty he received his master's degree in theoretical physics from Purdue University, and climbed the Matterhorn.

After contributing to the theoretical solution of the re-entry problem as an aerodynamicist at Bell Aircraft Corporation, Smith earned his doctorate in mathematics at Columbia University, subsequently embarking upon a 30-year career as a Professor of Mathematics at MIT, UCLA, and Oregon State University.

Above all, however, it needs to be realized that despite his impeccable credentials in physics, mathematics, and philosophy, Wolfgang Smith is at heart an outsider not only in regard to these academic disciplines, but more profoundly, in reference to the post-Enlightenment premises of our contemporary world. Early in life he became deeply attracted to the Platonist and Neoplatonist schools, and subsequently undertook extensive sojourns in India and the Himalayan regions to contact such vestiges of ancient tradition as still could be found. And one of the basic lessons he learned by way of these encounters is that there actually exist higher sciences in which man himself plays the part not merely of the observer, but of the "scientific instrument": becomes himself, in other words, the "microscope" or "telescope" by which he is enabled to perceive hitherto invisible reaches of the integral cosmos. By the same token, moreover, Smith came to recognize the stringent limitations to which our contemporary sciences are subject by virtue of their "extrinsic" *modus operandi*: the folly of presuming to fathom the depths of the universe having barely scratched the surface in the discovery of man himself.

Finding himself, thus, irreconcilably at odds with the prevailing Zeitgeist, Smith decided to forego a professional career in the fields of his primary interest—i.e., physics and philosophy—in favor of pure mathematics: the one and only academic discipline, he avers, in which "political correctness" can find no foothold. And so he enjoyed the luxury of pursuing a respected university career while being at liberty, as he puts it, "to remain perfectly sane."

It is no wonder, then, that when he finally confronted the so-called quantum enigma, Smith perceived the issue in a very different light than his peers. The problem all along had actually not been "technical"! It was not a question to be resolved by way of differential equations, nor primarily a matter of finding something new—but one of jettisoning an entire Weltanschauung. And for Wolfgang Smith this posed no difficulty: he had in fact done so decades earlier, as can be discerned in his remarkable series of publications.

Wolfgang Smith's life and work are the subject of the documentary film *The End of Quantum Reality*, scheduled for release in early 2019.

1657-20

CPSIA information can be obtained
at www.ICGtesting.com
Printed in the USA
LVHW111359240820
664065LV00001B/340

9 781621 384298